Selected Titles in This Series

685 **Jean-Luc Joly, Guy Metivier, and Jeffrey Rauch,** Caustics for dissipative semilinear oscillations, 2000

684 **Harvey I. Blau, Bangteng Xu, Z. Arad, E. Fisman, V. Miloslavsky, and M. Muzychuk,** Homogeneous integral table algebras of degree three: A trilogy, 2000

683 **Serge Bouc,** Non-additive exact functors and tensor induction for Mackey functors, 2000

682 **Martin Majewski,** Rational homotopical models and uniqueness, 2000

681 **David P. Blecher, Paul S. Muhly, and Vern I. Paulsen,** Categories of operator modules (Morita equivalence and projective modules), 2000

680 **Joachim Zacharias,** Continuous tensor products and Arveson's spectral C^*-algebras, 2000

679 **Y. A. Abramovich and A. K. Kitover,** Inverses of disjointness preserving operators, 2000

678 **Wilhelm Stannat,** The theory of generalized Dirichlet forms and its applications in analysis and stochastics, 1999

677 **Volodymyr V. Lyubashenko,** Squared Hopf algebras, 1999

676 **S. Strelitz,** Asymptotics for solutions of linear differential equations having turning points with applications, 1999

675 **Michael B. Marcus and Jay Rosen,** Renormalized self-intersection local times and Wick power chaos processes, 1999

674 **R. Lawther and D. M. Testerman,** A_1 subgroups of exceptional algebraic groups, 1999

673 **John Lott,** Diffeomorphisms and noncommutative analytic torsion, 1999

672 **Yael Karshon,** Periodic Hamiltonian flows on four dimensional manifolds, 1999

671 **Andrzej Rosłanowski and Saharon Shelah,** Norms on possibilities I: Forcing with trees and creatures, 1999

670 **Steve Jackson,** A computation of δ^1_5, 1999

669 **Seán Keel and James McKernan,** Rational curves on quasi-projective surfaces, 1999

668 **E. N. Dancer and P. Poláčik,** Realization of vector fields and dynamics of spatially homogeneous parabolic equations, 1999

667 **Ethan Akin,** Simplicial dynamical systems, 1999

666 **Mark Hovey and Neil P. Strickland,** Morava K-theories and localisation, 1999

665 **George Lawrence Ashline,** The defect relation of meromorphic maps on parabolic manifolds, 1999

664 **Xia Chen,** Limit theorems for functionals of ergodic Markov chains with general state space, 1999

663 **Ola Bratteli and Palle E. T. Jorgensen,** Iterated function systems and permutation representation of the Cuntz algebra, 1999

662 **B. H. Bowditch,** Treelike structures arising from continua and convergence groups, 1999

661 **J. P. C. Greenlees,** Rational S^1-equivariant stable homotopy theory, 1999

660 **Dale E. Alspach,** Tensor products and independent sums of $\mathcal{L}_p$-spaces, $1 < p < \infty$, 1999

659 **R. D. Nussbaum and S. M. Verduyn Lunel,** Generalizations of the Perron-Frobenius theorem for nonlinear maps, 1999

658 **Hasna Riahi,** Study of the critical points at infinity arising from the failure of the Palais-Smale condition for n-body type problems, 1999

657 **Richard F. Bass and Krzysztof Burdzy,** Cutting Brownian paths, 1999

656 **W. G. Bade, H. G. Dales, and Z. A. Lykova,** Algebraic and strong splittings of extensions of Banach algebras, 1999

655 **Yuval Z. Flicker,** Matching of orbital integrals on $GL(4)$ and $GSp(2)$, 1999

654 **Wancheng Sheng and Tong Zhang,** The Riemann problem for the transportation equations in gas dynamics, 1999

(Continued in the back of this publication)

Caustics for Dissipative Semilinear Oscillations

MEMOIRS
of the
American Mathematical Society

Number 685

Caustics for Dissipative Semilinear Oscillations

Jean-Luc Joly
Guy Metivier
Jeffrey Rauch

March 2000 • Volume 144 • Number 685 (third of 5 numbers) • ISSN 0065-9266

American Mathematical Society
Providence, Rhode Island

2000 *Mathematics Subject Classification.*
Primary 35L60, 35B05, 35G25.

Library of Congress Cataloging-in-Publication Data

Joly, Jean-Luc, 1942–
Caustics for dissipative semilinear oscillations / Jean-Luc Joly, Guy Metivier, Jeffrey Rauch.
p. cm. — (Memoirs of the American Mathematical Society, ISSN 0065-9266 ; no. 685)
"Volume 144, number 685 (third of 5 numbers)."
Includes bibliographical references.
ISBN 0-8218-2041-9 (alk. paper)
1. Differential equations, Hyperbolic—Numerical solutions. 2. Differential equations, Nonlinear—Numerical solutions. I. Métivier, Guy II. Rauch, Jeffrey. III. Title. IV. Series.
QA3 .A57 no. 685
[QA377]
510 s–dc21
[515′.353] 99-058329

Memoirs of the American Mathematical Society

This journal is devoted entirely to research in pure and applied mathematics.

Subscription information. The 2000 subscription begins with volume 143 and consists of six mailings, each containing one or more numbers. Subscription prices for 2000 are \$466 list, \$419 institutional member. A late charge of 10% of the subscription price will be imposed on orders received from nonmembers after January 1 of the subscription year. Subscribers outside the United States and India must pay a postage surcharge of \$30; subscribers in India must pay a postage surcharge of \$43. Expedited delivery to destinations in North America \$35; elsewhere \$130. Each number may be ordered separately; *please specify number* when ordering an individual number. For prices and titles of recently released numbers, see the New Publications sections of the *Notices of the American Mathematical Society.*

Back number information. For back issues see the *AMS Catalog of Publications.*

Subscriptions and orders should be addressed to the American Mathematical Society, P. O. Box 5904, Boston, MA 02206-5904. *All orders must be accompanied by payment.* Other correspondence should be addressed to Box 6248, Providence, RI 02940-6248.

Memoirs of the American Mathematical Society is published bimonthly (each volume consisting usually of more than one number) by the American Mathematical Society at 201 Charles Street, Providence, RI 02904-2294. Periodicals postage paid at Providence, RI. Postmaster: Send address changes to Memoirs, American Mathematical Society, P. O. Box 6248, Providence, RI 02940-6248.

This publication is indexed in *Science Citation Index*®, *SciSearch*®, *Research Alert*®, *CompuMath Citation Index*®, *Current Contents*®/*Physical, Chemical & Earth Sciences.*
Printed in the United States of America.

∞ The paper used in this book is acid-free and falls within the guidelines established to ensure permanence and durability.
Visit the AMS home page at URL: `http://www.ams.org/`

10 9 8 7 6 5 4 3 2 1 05 04 03 02 01 00

Contents

Abstract

This paper describes focusing effects near caustics for high frequency solutions of first order semilinear dissipative hyperbolic systems $Lu + F(u) = 0$ or analogous semilinear wave equations. The dissipitivity assumption is made because for general equations nonlinear interactions can amplify the growth of amplitudes of focusing solutions and lead to noncontinuation of solutions, as shown in [JMR 1] [JMR 2]. For the dissipative equations we consider, global solutions exist and we study their behavior beyond caustics. As in the linear theory, we use Lagangian integrals to describe the oscillations near the caustics. The nonlinear solutions are well approximated by oscillatory integrals on domains which may include caustics. The approximation is valid in L^p, for an exponent p related to the growth at infinity of the nonlinear dissipative terms. Away from the caustics, phase-amplitude expansions are recovered by the stationary phase theorem.

A key difficulty (and novelty) is to analyse the action of nonlinear functions on oscillatory integrals. This relies on new sharp uniform bounds for the L^p norm of oscillatory integrals. This analysis reveals a critical exponent p_c, associated to each point of the caustic. For generic caustic points, this exponent is 3. For radial focusing in $\mathbb{R}^d$, $d \geq 2$, it is $1 + 2/(d-1)$. When the nonlinearity $F(u)$ grows at infinity faster than $|u|^{p_c}$ the dissipative mechanism is strong and oscillations are absorbed at caustics so there are no oscillations past the focusing point. This extends the results in [JMR 2] from focusing spherical wavefronts to general caustics. When the nonlinear term grows at infinity at a strictly slower rate than $|u|^{p_c}$, oscillations cross caustics and amplitudes can be computed after focusing. This generalizes the results of [JMR 3] to superlinear dissipative equations.

Research partially supported by the U.S. National Science Foundation, U.S. Office of Naval Research, and the NSF-CNRS cooperation program under grants number NSF-DMS-9803296 and OD-G-N0014-92-J-1245 NSF-INT-9314095 respectively, and the CNRS through the Groupe de Recherche G1180 POAN.

Received by the editor February 26, 1998.

1. Introduction

This paper describes focusing effects near general caustics for high frequency solutions of semilinear dissipative equations. Several consequences of focusing mechanisms have already been described. In [JMR 1] and [JMR 2] it was shown that nonlinear interactions can amplify the growth of amplitudes due to focusing and lead to noncontinuation of solutions, even as weak solutions. For globally Lipschitzean or dissipative nonlinearities this phenomena does not occur. There are unique global solutions and one can pose the problem of describing the behavior of oscillations beyond caustics. In [JMR 2] we proved that strong dissipation can absorb radial oscillations. In [JMR 3] we showed how nonlinear oscillatory waves cross caustics when the nonlinearity is globally Lipschitzean. In the latter paper the solutions are described by oscillatory integrals which define bounded families in L^2 spaces.

The analysis in this paper relies on sharp uniform bounds for the L^p norm of such oscillatory integrals. Such estimates have already been obtained in [R 1] [R 2] [Sv]. However, the assumptions used in these papers are not adapted to our purpose. In section 3, we state and prove the uniform L^p estimates which we need. We use them to study a class of dissipative equations and consider general caustics. The analysis reveals a critical exponent p_c, associated to each point of the caustic set where rays focus. For generic caustic points, this exponent is 3. For radial focusing in $\mathbb{R}^d$, $d \geq 2$, it is $1 + 2/(d-1)$.

When the nonlinearity grows at infinity faster than $|u|^{p_c}$ the dissipative mechanism is strong and oscillations are absorbed at caustics so there are no oscillations past the focusing point. This extends the results in [JMR 2] from focusing spherical wavefronts to general caustics. When the nonlinear term grows at infinity at a strictly slower rate than $|u|^{p_c}$, oscillations cross caustics and amplitudes can be computed after focusing. This generalizes the results of [JMR 3] to superlinear dissipative equations.

We now sketch the results for a typical example. Consider the semilinear dissipative wave equation

$$\Box u + |\partial_t u|^{p-1}\, \partial_t u = 0 \tag{1.1}$$

with oscillatory Cauchy data

$$u^\varepsilon\big|_{t=0} = \underline{u}_0(x) + \varepsilon\, \mathbf{U}_0(x, \psi(x)/\varepsilon)\,, \qquad \partial_t u^\varepsilon\big|_{t=0} = \underline{u}_1(x) + \mathbf{U}_1(x, \psi(x)/\varepsilon)\,, \tag{1.2}$$

where ψ is a smooth function with nonvanishing differential on $\omega \subset \mathbb{R}^d$, $\mathbf{U}_0(x,\theta)$ and $\mathbf{U}_1(x,\theta)$ are smooth, 2π-periodic in θ with mean 0 and compactly supported in $\omega \times \mathbb{T}$ and $\underline{u}_0, \underline{u}_1 \in C_0^\infty(\mathbb{R}^d)$. Later, these regularity assumptions are considerably weakened.

Since the Cauchy data (1.2) define bounded families in $H^1(\mathbb{R}^d)$ and $L^2(\mathbb{R}^d)$, the corresponding weak solutions u^ε are bounded in $C^0([0,+\infty[;H^1(\mathbb{R}^d))$ with $\partial_t u^\varepsilon$ bounded $C^0([0,+\infty[;L^2(\mathbb{R}^d)) \cap L^{p+1}([0,+\infty[\times\mathbb{R}^d)$, see [L], [LS], [Str].

The behavior of u^ε for small times is given by [JR]. The solution satisfies

$$(1.3) \qquad u^\varepsilon \;=\; \underline{u} + \varepsilon\,\mathbf{U}_+(t,x,\varphi_+(t,x)/\varepsilon) \;+\; \varepsilon\,\mathbf{U}_-(t,x,\varphi_-(t,x)/\varepsilon) \;+\; \varepsilon^2 \dots ,$$

where $\varphi_\pm$ are the solutions of the eikonal equation for $\Box$ with data ψ at $t=0$ and the profiles $\mathbf{U}_\pm$ satisfy transport equations, coupled to a nonlinear wave equation for $\underline{u}$.

In general, $\varphi_\pm$ develop singularities in finite time. The singular locus is called the caustic set $\mathcal{C}$. As one approaches $\mathcal{C}$, amplitudes tend to infinity and the description (1.3) breaks down. For linear equations, the substitute for (1.3) is well known. The phases are replaced by Lagrangian manifolds $\Lambda_\pm$ which are globally defined and smooth and the oscillations $\mathbf{U}_\pm(\,\cdot\,,\varphi_\pm/\varepsilon)$ are replaced by Lagrangian distributions associated with $\Lambda_\pm$ (see e.g. [Du],[DH], [Hö 1], [Hö 2]). This representation shows that after $\mathcal{C}$ more phases are required to describe the asymptotics of u^ε. In the nonlinear case the situation can be worse since amplitudes may blow up before reaching the caustics [JMR 2]. This does not happen for dissipative equations.

For the wave equation, the Lagrangian distributions are given by integrals of the form

$$(1.4) \qquad (2\pi\varepsilon)^{-d} \iint e^{i\Phi_\pm(t,x,y,\xi)/\varepsilon}\, a(t,y)\, dy\, d\xi\,,$$

with

$$(1.5) \qquad \Phi_\pm(t,x,y,\xi) \;:=\; \pm t\,|\xi| \;+\; (x-y)\cdot\xi \;+\; \psi(y)\,.$$

In the nonlinear case one takes care of harmonics by considering symbols depending on the extra variable $\theta \in \mathbb{T} := \mathbb{R}/2\pi\mathbb{Z}$. The periodic symbols with mean 0

$$(1.6) \qquad \mathcal{A}(t,y,\theta) \;=\; \sum_{n\neq 0} a_n(t,y)\; e^{i\,n\,\theta}$$

yield oscillatory integrals

$$(1.7) \qquad I^\varepsilon_\pm(\mathcal{A}) \;=\; \sum_{n\neq 0} \frac{|n|^d}{(2\pi\varepsilon)^d} \iint e^{i\,n\,\Phi_\pm(t,x,y,\xi)/\varepsilon}\; a_n(t,y)\; dy\, d\xi\,.$$

The linear theory (see e.g. [Du], [Hö 2]), shows that

$$(1.8) \qquad u^\varepsilon \;\sim\; \underline{u} \;+\; \varepsilon\, I^\varepsilon_+(\mathcal{U}_+) \;+\; \varepsilon\, I^\varepsilon_-(\mathcal{U}_-)\,.$$

The caustic set $\mathcal{C}$, is the set of points (t,x) where at least one of the critical points of $(y,\xi) \mapsto \Phi_\pm(t,x,y,\xi)$ is degenerate. Outside the caustic set, the classical stationary phase theorem yields a phase-amplitude expansion of the form

$$(1.9) \qquad I^\varepsilon_\pm(\mathcal{U}) \;=\; \sum_k \mathbf{U}_{\pm,k}(t,x,\varphi_{\pm,k}(t,x)/\varepsilon) \;+\; O(\varepsilon)\,.$$

The sum (1.9) runs over the set of critical points of $(y,\xi) \mapsto \Phi_\pm(t,x,y,\xi)$. In particular, for small times, there is a unique critical point and (1.8) is identical to (1.3). Sometimes refined stationary phase theorems yield precise descriptions of u^ε near the caustic, [Lu], [Du], [Hö 2].

In [JMR 3], Lagrangian distributions are at the center of the analysis for the nonlinearities which are globally Lipschitzean. The goal of this paper, is to show that the description (1.8) is true for all dissipative equations (1.1), provided that the Lagrangians $\Lambda_\pm$ satisfy a nonresonance condition and some technical assumptions.

The analysis in [JMR 3] uses the fact that families of integrals like (1.4) or (1.7) are bounded in L^2, even on neighborhoods of caustics. This expresses the boundedness of energy. Near the caustics, these families are not bounded in L^∞. It is a natural question, which is essential to control the nonlinear terms, to determine the supremum of the real numbers $q \geq 2$ such that $I_\pm^\varepsilon(\mathcal{A})$ is bounded in L^q.

The analysis reveals a critical exponent

$$q_c := 2 + 2/\mu$$

which does not depend on the full complexity of the caustic but, for the phases $\Phi_\pm$ in (1.5), only on the algebraic order μ to which the determinant of the Hessian matrix of $\Phi_\pm$ vanishes at degenerate critical points.

When the C^∞ symbol $\mathcal{A}$ is supported in a small neighborhood of a degenerate critical point of multiplicity μ, then $I_\pm^\varepsilon(\mathcal{A})$ is bounded in the weak L^{q_c} space, i.e the Lorentz space $L^{q_c,\infty}$. Thus, for all $q < q_c$, $I_\pm^\varepsilon(\mathcal{A})$ is bounded in L^q.

This is proved under a locally constant rank condition for degenerate critical points. We refer to §2 and §3 for a precise statement. Similar estimates can be found in [R 1], [R 2] and [Sv], where additional convexity hypotheses are imposed. The critical index q_c is sharp. If $\mathcal{A} \neq 0$ at the degenerate critical points, then for $q \geq q_c$, the L^q norms of $I_\pm^\varepsilon(\mathcal{A})$ are not bounded. To guarantee their boundedness in L^q requires some vanishing of $\mathcal{A}$ on $\mathcal{S}_\pm$ the set of degenerate critical points of $\Phi_\pm$. Results in this direction can be found in [CDMM] (see also references in [Ste]). Note that the condition $q < q_c$ is also necessary for the right hand side of (1.9) to define bounded families in L^q.

These estimates are not and do not follow the from sharp L^p estimates on Fourier Integral Operators which can be found in [Ste] and [So] and references therein.

We show that

There are $\underline{u}$ and profiles $\mathcal{U}_+$ and $\mathcal{U}_-$ such that the solutions of (1.1) (1.2) satisfy (1.8). $\underline{u}$ and $\mathcal{A}_\pm$ are related by a coupled system of transport equations and a wave equation. With $\mathcal{B}_\pm := \partial_\theta \mathcal{U}_\pm$, the equations are of the form :

$$\partial_t \mathcal{B}_\pm + \mathcal{E}_\pm(\underline{u}, \mathcal{B}_+, \mathcal{B}_-) = 0 \quad , \quad \Box \underline{u} + \underline{\mathcal{E}}(\underline{u}, \mathcal{B}_+, \mathcal{B}_-) = 0 . \tag{1.10}$$

The relation (1.9) gives a phase-amplitude representation of u^ε outside $\mathcal{C}$. The propagation equation for the amplitude $\mathbf{U}_{\pm,j}$ follows from the equation for $\mathcal{U}_\pm$. The $\mathbf{U}_{\pm,j}$ satisfy transmission conditions which include the usual phase shift across caustics, as in [JMR 3]. The system (1.10) inherits dissipativity from (1.1). The nonlinear interactions $\mathcal{E}_\pm$ are singular above the caustic set $\mathcal{C}$. This fact was exploited in [JMR 3] to reveal the possible occurrence of nonsmooth transfer of energy at caustics. Here, the singularity of $\mathcal{E}_\pm$ is augmented by the stronger nonlinearity in (1.1), and $\mathcal{E}_\pm(\underline{u}, \mathcal{B}_+, \mathcal{B}_-)$ is not always locally integrable near the caustic points. Thus the transport equations in (1.10) must be interpreted with care. The discussion again involves the critical index q_c attached to degenerate critical points, indicating consistency with the property that $\partial_t u^\varepsilon$ belongs to L^{p+1}. If $p+1 \geq q_c$, then $\mathcal{E}_\pm(\underline{u};\mathcal{B}_+,\mathcal{B}_-)$ is not necessarily integrable near the corresponding point. But there, *strong* dissipation and *absorption* occur. The transport equation (1.10) is satisfied along the rays before the singular point, $\mathcal{B}_\pm$ tends to zero at this point, and $\mathcal{B}_\pm$ vanishes after it. On the contrary, when $p+1 < q_c$, then $\mathcal{E}_\pm(\underline{u};\mathcal{B}_+,\mathcal{B}_-)$ is locally integrable so that the transport equation in (1.10) makes sense and is satisfied across the singular point.

Denote by $T_\pm(y)$, the smallest focusing time t along the ray $y \pm t\, d\psi(y)/|d\psi(y)|$ and such that the critical exponent q_c at (t,y) is less than or equal to $p+1$. For the wave equation, the times of focusing along the rays from $\underline{y}$ are $\pm$ the radii of curvature of the level surface $\psi(y) = \psi(\underline{y})$. The multiplicity $\mu_\pm(t,\underline{y})$ is the multiplicity of the corresponding principal curvature. Then the precise meaning of (1.10) is the following.

$\mathcal{B}_\pm = 0$ for $t > T_\pm(y)$ and the transport equations $\partial_t \mathcal{B}_\pm = \mathcal{E}_\pm(\underline{u},\mathcal{B}_+,\mathcal{B}_-)$ are satisfied in L^1_{loc} for $t < T_\pm(y)$.

When $p \geq 3$, then $p+1$ is always larger than $2 + 2/\mu$ since $\mu \geq 1$. Thus $T_\pm(y)$ is the first time of focusing along the ray from y.

For generic caustics, the multiplicity $\mu = 1$ and the critical index $q_c = 4$. Thus, if (almost) all the rays have only generic focusing points, absorption occurs at caustics points if and only if $p \geq 3$, which is the supercritical case. On the other hand, in the subcritical case, that is when $p < 3$, oscillations cross the caustics (Figure 1.1 below).

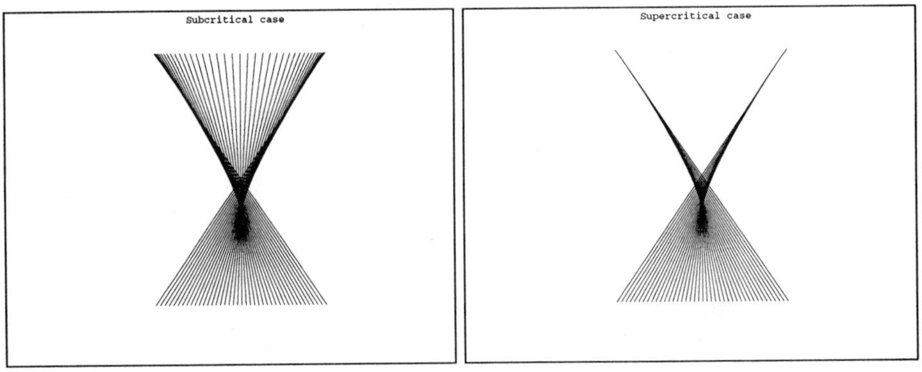

Figure 1.1

For radial phases $\psi(y) = \chi(|y|)$, the multiplicity μ is equal to $d-1$. Thus absorption occurs at the caustic set $\{x = 0\}$ if and only if $p \geq 1 + 2/(d-1)$. For such radial phases, and for $p \geq 1 + 2/(d-1)$, absorption was proved in [JMR 2]. The present paper improves the asymptotics near the caustic, shows that the index $1 + 2/(d-1)$ is sharp and that, for smaller p, oscillations continue after focusing.

When $p < 1 + 2/(d-1)$, absorption never occurs and oscillations persist after crossing the caustics, since the multiplicity μ is at most equal to $d-1$. This includes the limit case, $p = 1$ which corresponds to linear equations (1.1) and also to globally Lipschitzean nonlinearities.

2. Notations and main results

2.1. The time evolution

In $\mathbb{R}^{1+d}$, consider the Cauchy problem for a semilinear system

$$\partial_t u - \sum_{j=1}^{d} A_j \,\partial_j u \,+\, f(t,x,u) \;=\; 0\,, \quad u\big|_{t=0} = u_0 \tag{2.1.1}$$

where $\partial_j := \partial/\partial x_j$. $L := \partial_t - \sum_{j=1}^d A_j\,\partial_j$ is assumed to be a symmetric hyperbolic first order system of constant multiplicity. For simplicity, we assume that it has constant coefficients.

ASSUMPTION 2.1.1. *The A_j are constant real symmetric $N \times N$ matrices and the eigenvalues $\lambda_k(\xi)$, $1 \leq k \leq k_0$, of $A(\xi) := \sum_{j=1}^d \xi_j A_j$ have multiplicity independent of $\xi \in \mathbb{R}^d \backslash 0$.*

The nonlinear interaction f is assumed to be *dissipative.*

ASSUMPTION 2.1.2. *f is a C^1 function from $[0,+\infty[\times\mathbb{R}^d \times \mathbb{C}^N$ to $\mathbb{C}^N$, with $f(t,x,0) = 0$ and there are $p \in [1,+\infty[$ and constants $0 < c < C < +\infty$ such that for all $(t,x) \in \mathbb{R}^{1+d}$ and all u and v in $\mathbb{C}^N$,*

$$|\,F(t,x,u)\,| \leq \; C\,|\,u\,|^p \quad , \quad |\,\partial_{u,\overline{u}} f(t,x,u)\,| \leq \; C\,|\,u\,|^{p-1}\,, \tag{2.1.2}$$

$$\mathrm{Re}\,\Big(\big(f(t,x,u) - f(t,x,v)\big)\cdot(\overline{u} - \overline{v})\Big) \;\geq\; c\,|\,u\,-\,v\,|^{p+1}\,. \tag{2.1.3}$$

In (2.1.2), $\partial_{u,\overline{u}} f$ denotes the set of derivatives $\partial_u f$ and $\partial_{\overline{u}} f$. Remark that $f(t,x,0) = 0$ and (2.1.3) imply that

$$c\,|\,u\,|^{p+1} \;\leq\; \mathrm{Re}\, f(t,x,u)\cdot\overline{u}\,. \tag{2.1.4}$$

EXAMPLE. $f := \partial_{\overline{u}}\Phi : \mathbb{C}^N \to \mathbb{C}^N$ satisfies Re $\Big(\big(f(u) - f(v)\big)\cdot(\overline{u} - \overline{v})\Big) \geq 0$, when Φ is a convex function from $\mathbb{C}^N$ to $\mathbb{R}$. The stronger condition (2.1.3), and the other assumptions are satisfied for example for $\Phi(u) := |u|^{p+1}$.

In most applications, u is real valued and f is defined for real values of u, but the use of complex exponentials in place of sin and cos is convenient and we systematically introduce complex values for u. This is not a restriction, since when f is a C^1 function from $[0,+\infty[\times\mathbb{R}^d\times\mathbb{R}^N$ to $\mathbb{R}^N$ which satisfies the analogues of (2.1.2)-(2.1.3) for u and v in $\mathbb{R}^N$, the extension to complex $u+iu'\in\mathbb{C}^N$, defined by $\widetilde{f}(u+iu'):=f(u)+i\,g(u')$, satisfies (2.1.2) (2.1.3) if g has the same properties as f.

Assumption 2.1.2 can be slightly weakened. In particular, in §7, we give modifications of (2.1.2)-(2.1.3) adapted to the first order reduction of the wave equation (1.1).

Solutions of the Cauchy problem (2.1.1) are constructed in [Str] (see also [LS] or [Ha]).

THEOREM 2.1.3. *For $u_0\in L^2(\mathbb{R}^d)$, the Cauchy problem (2.1.1) has a unique solution $u\in C^0(\mathbb{R};L^2(\mathbb{R}^d))\cap L^{p+1}([0,+\infty[\times\mathbb{R}^d)$. Moreover, if u_0 and v_0 are Cauchy data in $L^2(\mathbb{R}^d)$, the solutions u and v satisfy*

$$(2.1.5)\quad \sup_{t\geq 0}\,\|\,u(t)-v(t)\,\|^2_{L^2(\mathbb{R}^d)}\;+\;2c\,\|\,u-v\,\|^{p+1}_{L^{p+1}([0,+\infty[\times\mathbb{R}^d)}\;\leq\;\|\,u_0-v_0\,\|^2_{L^2(\mathbb{R}^d)}\,.$$

2.2. Lagrangians, caustics and phases

Consider Cauchy data of the form

$$(2.2.1)\qquad u_0^\varepsilon(x)\;=\;\underline{u}_0(x)\;+\;\mathcal{U}_0(x,\psi(x)/\varepsilon),$$

where $\mathcal{U}_0(x,\theta)$ is periodic in θ with vanishing mean value.

ASSUMPTION 2.2.1. *ω is an open subset of $\mathbb{R}^d$ and $\partial\omega$ has d-dimensional Lebesgue measure equal to zero. The initial phase ψ is real valued and C^∞ on a neighborhood of $\overline{\omega}$ and $d\psi$ does not vanish on $\overline{\omega}$.*

Extensions to phases with isolated critical points or singular points are possible, as in [JMR 3] §6.5.

We assume that the profile $\mathcal{U}$ is defined on $\mathbb{R}^d\times S^1$ with $\mathcal{U}_0(x,\theta)=0$ when $x\notin\omega$. Then the Cauchy data (2.2.1) is well defined for $x\in\mathbb{R}^d$.

Thanks to Assumption 2.1.1, the eiconal equation $\det L(d\varphi)=0$ is equivalent to

$$(2.2.2)\qquad \partial_t\varphi\;-\;\lambda_k(\partial_x\varphi)=0\,,\quad \varphi_{|t=0}=\psi\,,$$

for some $k\in\{1,\dots,k_0\}$. The local solutions of (2.2.2) are associated to global Lagrangian manifolds Λ_k in $T^*\mathbb{R}^{1+d}$. As in [JMR 3], introduce global coordinates (t,y) on theses manifolds. Precisely, define

$$(2.2.3)\qquad G\;:=\;[0,+\infty[\times\omega\,,$$

and the mappings $\jmath_k\,:\,\overline{G}=[0,+\infty[\times\overline{\omega}\;\rightarrow\;[0,+\infty[\times\mathbb{R}^d$

$$(2.2.4)\qquad \jmath_k(t,y)=(t,x)\quad\text{with}\quad x=q_k(t,y):=y-t\lambda_k'(d\psi(y))\,.$$

Introduce the set of points where $\jmath_k$ is not a local diffeomorphism,

$$\mathcal{S}_k := \{ (t,y) \in G \quad | \quad \det \partial_y q_k(t,y) = 0 \} . \tag{2.2.5}$$

The caustic set $\mathcal{C}_k$ is

$$\mathcal{C}_k := \jmath_k(\mathcal{S}_k) \subset [0,+\infty[\times\mathbb{R}^d . \tag{2.2.6}$$

Introduce

$$g_k(y) := \lambda_k'(d\psi(y)) \tag{2.2.7}$$

and its Jacobian matrix $g_k'(y)$. Then $(t,y) \in \mathcal{S}_k$ if and only if $1/t$ is a positive eigenvalue of $g_k'(y)$.

The analysis of the singularities near the caustic set depends on the properties of the mappings $\jmath_k$. We make the following *finiteness* hypothesis.

ASSUMPTION 2.2.2. *There is an integer ℓ_*, such that for all k, and all $(t,x) \notin \mathcal{C}_k$, the number of points $y \in \omega$ such that $x = q_k(t,y)$, is less than or equal to ℓ_*.*

EXAMPLE. The hypothesis is satisfied when ψ is real analytic since in that case λ is also real analytic.

As in [JMR 3], introduce the following terminology

DEFINITION 2.2.3. *A subset $\Omega \subset [0,+\infty[\times\mathbb{R}^d$ is k-regular if and only if it has the following properties.*

i) Ω is a connected relatively open subset of $[0,+\infty[\times\mathbb{R}^d$.

ii) There are smooth functions $\rho_{k,1},..,\rho_{k,\ell}$ from Ω to ω such that $\rho_{k,j} \neq \rho_{k,l}$ for $j \neq l$ and

$$\forall (t,x) \in \Omega , \quad \{ y \in \omega \mid q_k(t,y) = x \} = \{ \rho_{k,j}(t,x) \}_{1 \le j \le \ell} . \tag{2.2.8}$$

A subset $\Omega \subset [0,+\infty[\times\mathbb{R}^d$ is called regular, when it is k-regular for all k.

Note that the $\rho_{k,j}(t,x)$ are the feet of the bicharacteristics passing through (t,x). In *ii)* the number ℓ is allowed to be zero. In this case, condition *ii)* means that $\{ y \in \omega \mid q_k(t,y) = x \}$ is empty. A k-regular set is necessarily contained in $[0,+\infty[\times\mathbb{R}^d \backslash \mathcal{C}_k$. As noticed in [JMR 3], the converse is not true, since the number of preimages of x can change when one of them reach the boundary $\partial\omega$. However one has the following result (see §4 in [JMR 3]).

PROPOSITION 2.2.4. *Introduce*

$$\widetilde{\mathcal{C}}_k := \mathcal{C}_k \cup \jmath_k([0,+\infty[\times\partial\omega) , \quad \widetilde{\mathcal{C}} = \cup_k \widetilde{\mathcal{C}}_k . \tag{2.2.9}$$

Then $\widetilde{\mathcal{C}} \subset [0,+\infty[\times\mathbb{R}^d$ and, for each k, $\jmath_k^{-1}(\widetilde{\mathcal{C}}) \subset \overline{G}$ has zero Lebesgue measure. For all $(t,x) \notin \widetilde{\mathcal{C}}_k$, $\jmath_k^{-1}(\{t,x\})$ is contained in G. The number of points in $\jmath_k^{-1}(\{t,x\})$ is locally constant on $[0,+\infty[\times\mathbb{R}^d \backslash \widetilde{\mathcal{C}}_k$. In particular, every point $(t,x) \notin \widetilde{\mathcal{C}}_k$ has a k-regular neighborhood.

If Ω is a k-regular open set, $\jmath_k^{-1}(\Omega)$ is the disjoint union of open sets $G_{k,j} \subset G$ such that $\jmath_k$ is a diffeomorphism from $G_{k,j}$ onto Ω. The inverse mappings are $(t,x) \to (t, \rho_{k,j}(t,x))$ with $\rho_{k,j}$ from (2.2.8). This defines a family of phases $\varphi_{k,j}$ on Ω, $(k,j) \in Z$,

$$\varphi_{k,j}(t,x) := \psi(\rho_{k,j}(t,x)) . \tag{2.2.10}$$

The $\varphi_{k,j}$ satisfy the eiconal equation (2.2.2). They are used to describe the oscillations of the solutions of the linear equation $Lu^\varepsilon = 0$ with Cauchy data (2.2.1). These oscillations may interact nonlinearly, creating new oscillations for the solutions of (2.1.1) (see e.g. [McLPT], [HMR], [JMR 4] and references therein, for several studies on the mechanisms of nonlinear resonance of oscillations). In this paper, which is mainly concerned with caustics, we assume that the phases $\varphi_{k,j}$ do not resonate strongly. The reader is referred to [JMR 3] where it is shown that this assumption is satisfied by a wide class of examples.

ASSUMPTION 2.2.5. *For all regular open Ω the set of phases $\varphi_{k,j}$, $(k,j) \in Z$, given by (2.2.10), satisfy the following nonresonance property. For all $\{\alpha_{k,j}\} \in \mathbb{Z}^Z$ such that at least two components do not vanish,*

$$\det L\big(\sum \alpha_{k,j}\, d\varphi_{k,j}\big) \neq 0 \quad \text{a.e. on } \Omega . \tag{2.2.11}$$

2.3. L^p estimates of oscillatory integrals

Our goal is to represent the solutions u^ε of equation (2.1.1) with Cauchy data (2.2.1), as a sum of Lagrangian distributions associated to Λ_k. Let $\mathbb{T} = \mathbb{R}/2\pi\mathbb{Z}$. On $\mathbb{T}$ introduce the measure $d\theta$ with total mass equal to 1. Each term is given by an oscillatory integral with a symbol $\mathcal{U}(x,\theta)$ defined on $G \times \mathbb{T}$ with mean 0 in θ. The nonlinear interaction of harmonics is expressed by the coupling among the Fourier modes of $\mathcal{U}$ in the variable θ.

DEFINITION 2.3.1. *A trigonometric polynomial is a finite sum*

$$\mathcal{A}(t,y,\theta) := \sum_{n\neq 0} a_n(t,y)\, e^{in\theta} , \quad a_n \in C_0^\infty([0,+\infty[\times\omega) \tag{2.3.1}$$

The space of trigonometric polynomials is denoted by $\mathcal{P}$.

For $\mathcal{A} \in \mathcal{P}$, introduce the oscillatory integral

$$I_k^\varepsilon(\mathcal{A})\,(t,x) := \sum_n (|n|/2\pi\varepsilon)^d \int e^{i\,n\,\Phi_k(t,x,y,\xi)/\varepsilon}\, a_n(t,y)\, dy\, d\xi , \tag{2.3.2}$$

with

$$\Phi_k(t,x,y,\xi) := t\,\lambda_k(\xi) + (x-y)\cdot\xi + \psi(y) . \tag{2.3.3}$$

A crucial point in this work is to study the asymptotic behavior of such integrals, in L^q spaces. The boundedness in L^2 is clear, as well as the blow up in L^∞ near caustics. The question is to determine the set of q such that $I_k^\varepsilon(\mathcal{A})$ is bounded in L^q. The analysis reveals a critical exponent, which surprisingly does not depend on the full complexity of the caustic but only on the multiplicity μ_k.

DEFINITION 2.3.2. *The* **multiplicity**, *denoted* $\mu_k(t,y)$, *of* $\mathcal{S}_k$ *at a point* (t,y) *is the algebraic multiplicity of* $1/t$ *as an eigenvalue of* $g_k'(y)$. *The* **geometric multiplicity**, *denoted* $\nu_k(t,y)$, *that is the dimension of* $\ker(Id - t g_k'(y))$.

ASSUMPTION 2.3.3. *There is an open subset* $\omega_0 \subset \omega$ *such that* $\omega \backslash \omega_0$ *has Lebesgue measure equal to 0 and for all* k *and all* $(t,y) \in \mathcal{S}_k$ *with* $y \in \omega_0$, *one has*

i) μ_k *and* ν_k *are constant on a neighborhood of* (t,y) *in* $\mathcal{S}_k$, *and*

ii) either $\nu_k = 1$ *or* $\nu_k = \mu_k$.

EXAMPLES. 1. The assumption is satisfied for the wave equation or Maxwell equation when ψ is real analytic. 2. The assumption implies that on a neighborhood of $(t,y) \in \mathcal{S}_k$ with $y \in \omega_0$, $\mathcal{S}_k$ is a smooth manifold of codimension one.

DEFINITION 2.3.4. *For* $q \in [2, +\infty]$, $\mathcal{S}_k^-(q)$ *denotes the set of points* $(t,y) \in \mathcal{S}_k$ *such that* $y \in \omega_0$ *and* $\mu_k(t,y) < 2/(q-2)$. *The remaining set is denoted by* $\mathcal{S}_k^+(q) := \mathcal{S}_k \backslash \mathcal{S}_k^-(q)$.

The basic L^q estimate is given in the next Theorem.

THEOREM 2.3.5. *Suppose that Assumptions 2.2.2 and 2.3.3 hold, that* $q \geq 2$, *and* $\mathcal{A} \in \mathcal{P}$. *Suppose that the coefficients* a_n *vanish on a neighborhood of the points* $(t,y) \in \mathcal{S}_k^+(q)$. *Then,*

$$\sup_{\varepsilon \in]0,1]} \| \, I_k^\varepsilon(\mathcal{A}) \, \|_{L^q(\mathbb{R}^{1+d})} \; < \; +\infty \, . \tag{2.3.4}$$

REMARK 2.3.6. The limit $\mu_k < 2/(q-2)$ is sharp. If $y \in \omega_0$ and $\mathcal{A}$ does not vanish at a point $(t,y) \in \mathcal{S}_k^+$ where $\mu_k \geq 2/(q-2)$, then $I_k^\varepsilon(\mathcal{A})$ is not bounded in L^q (see §3).

2.4. Oscillations and profiles

When $(t,x) \notin \mathcal{C}_k$ or when the coefficients a_n vanish on a neighborhood of $\mathcal{S}_k$, the relevant stationary points with respect to (y,ξ) of the phase Φ_k, are nondegenerate. Therefore, the standard stationary phase expansion implies that

$$I_k^\varepsilon(\mathcal{A}) = J_k^\varepsilon(\mathcal{A}) \; + \; O(\varepsilon) \tag{2.4.1}$$

where

$$\begin{aligned} J_k^\varepsilon(\mathcal{A})\,(t,x) \; &:= \sum_{\{y \,|\, q_k(t,y)=x\}} \frac{1}{\Delta_k(t,y)} \Big(\sum_{n>0} i^{m_k(t,y)} \; a_n(t,y) \; e^{i\,n\,\psi(y)/\varepsilon} \\ &\qquad + \sum_{n<0} i^{-m_k(t,y)} \; a_n(t,y) \; e^{i\,n\,\psi(y)/\varepsilon} \Big) \\ &= \sum_{\{y \,|\, q_k(t,y)=x\}} \frac{1}{\Delta_k(t,y)} \; (\mathcal{H}^{m_k(t,y)} \mathcal{A})\,(t, y, \psi(y)/\varepsilon)\,, \end{aligned} \tag{2.4.2}$$

with

$$\Delta_k(t,y) := |\det d_y\, q_k(t,y)\,|^{1/2}\,, \tag{2.4.3}$$

$$m_k(t,y) := \frac{1}{2}\ \text{sign}\ \partial^2_{y,\xi}\Phi_k(t,x,y,d\psi(y))\,, \text{with } q_k(t,y)=x\,, \tag{2.4.4}$$

and $\mathcal{H}$ is the Hilbert transform acting on Fourier series by

$$\mathcal{H}\Big(\sum_{n\neq 0} a_n\ e^{in\theta}\Big)\ :=\ \sum_{n\neq 0} i^{\text{sign}\, n}\ a_n\ e^{in\theta}\,. \tag{2.4.5}$$

In (2.4.4), $\text{sign}\, A$ denotes the signature of the symmetric matrix A. Note that $\Delta_k(t,y)\neq 0$ for $(t,y)\notin \mathcal{S}_k$ and that $m_k(t,y)$ is locally constant on $G\backslash\mathcal{S}_k$.

Suppose that Ω is a regular open set. The matrix $\partial^2_{y,\xi}\Phi_k(t,x,y,d\psi(y))$ is nondegenerate for $(t,y)\notin\mathcal{S}_k$ and $x=q_k(t,y)$. Since Ω is connected, the signature $m_k(t,y)$ is constant on each component $G_{k,j}$ of $\jmath_k^{-1}(\Omega)$. Call $m_{k,j}$ its value. Then, using the notations (2.2.8) and (2.2.9), one has on Ω

$$J_k^\varepsilon(\mathcal{A})(t,x)\ =\ \sum_{j=1}^{\ell_k} \mathbf{A}_{k,j}\,(t,\, x,\, \varphi_{k,j}(t,x)/\varepsilon) \tag{2.4.6}$$

with

$$\mathbf{A}_{k,j}(t,x,\theta)\ :=\ \Delta_k(t,\rho_{k,j}(t,x))^{-1}\,(\mathcal{H}^{m_{k,j}}\mathcal{A})\ (t,\rho_{k,j}(t,x),\theta)\,. \tag{2.4.7}$$

This formula links the profile $\mathcal{A}$ on the Lagrangian and the profile $\mathbf{A}$ on the base. Note that Δ_k vanishes exactly to order $\mu_k/2$ on $\mathcal{S}_k$. Using the nonresonance Assumption 2.2.5, Proposition 2.4.2 shows that the L^q norm of $J_k^\varepsilon(\mathcal{A})$ is asymptotically estimated to be the L^q norm of the profiles $\mathbf{A}_{k,j}$. Since $\jmath_k$ is a diffeomorphism from $G_{k,j}$ to Ω whose Jacobian is Δ^2, the L^q norm of $\mathbf{A}_{k,j}$ is equal to a weighted L^q norm of $\mathcal{A}$ on $G_{k,j}\times\mathbb{T}$. This suggests the following definition.

DEFINITION 2.4.1. *For $q\in]1,+\infty[$, $\mathcal{L}^q_k$ denotes the space of L^q integrable functions on $[0,+\infty[\times\omega\times\mathbb{T}$ with respect to the measure $\Delta_k(t,y)^{2-q}\,dt\,dy\,d\theta$. For $0<s<1$, $\mathcal{L}^{s,q}_k$ denotes the space of functions $\mathcal{U}\in\mathcal{L}^q_k$ such that*

$$\int_0^1\int_{[0,T]\times\omega\times\mathbb{T}} \Delta_k(t,y)^{2-q}\ \Big|\frac{\mathcal{U}_k(t,y,\theta+h)\ -\ \mathcal{U}_k(t,y,\theta)}{h^s}\Big|^q dt\,dy\,d\theta\,\frac{dh}{h}\ <\ \infty\,.$$

For a trigonometric polynomial, belonging to $\mathcal{L}^q_k$ requires no condition on $\mathcal{S}^-_k(q)$ but some vanishing on $\mathcal{S}^+_k(q)$. $\mathcal{L}^{s,q}_k$ is the space of L^q integrable functions on $[0,+\infty[\times\omega$ with respect to the measure $\Delta_k(t,y)^{2-q}dt\,dy$ with values in the Sobolev space $W^{s,q}(\mathbb{T})$. Recall that

$$\|w\|^q_{W^{s,q}(\mathbb{T})}\ =\ \|w\|^q_{L^q(\mathbb{T})}\ +\ \int_0^1\int_{\mathbb{T}}\Big|\frac{w(\theta+h)\ -\ w(\theta)}{h^s}\Big|^q d\theta\,\frac{dh}{h}\,.$$

When $q=2$, note that $\mathcal{L}^2_k = L^2([0,T]\times\omega\times\mathbb{T})$.

PROPOSITION 2.4.2. *i) For all $q \in]1,\infty[$, there is a constant $C > 0$ such that for all trigonometric polynomials $\mathcal{A}$ which vanish on a neighborhood of $\mathcal{S}_k$ and all $T \geq 0$,*

$$\begin{aligned}\frac{1}{C}\,\|\mathcal{A}\|_{\mathcal{L}^q_k([0,T]\times\omega\times\mathbb{T})} \;&\leq\; \liminf_{\varepsilon\to 0}\,\|J^\varepsilon_k(\mathcal{A})\|_{L^q([0,T]\times\mathbb{R}^d)}\\ &\leq \limsup_{\varepsilon\to 0}\,\|J^\varepsilon_k(\mathcal{A})\|_{L^q([0,T]\times\mathbb{R}^d)} \;\leq\; C\,\|\mathcal{A}\|_{\mathcal{L}^q_k([0,T]\times\omega\times\mathbb{T})}.\end{aligned} \tag{2.4.8}$$

ii) For $1/q < s < 1$ and each $\varepsilon \in]0,1]$, J^ε_k extends to a bounded linear operator from $\mathcal{L}^{s,q}_k$ to $L^q([0,+\infty[\times\mathbb{R}^d)$ and

$$\sup_{\varepsilon\in]0,1]}\,\|J^\varepsilon_k(\mathcal{A})\|_{L^q([0,T]\times\mathbb{R}^d)} \;\leq\; C\,\|\mathcal{A}\|_{\mathcal{L}^{s,q}_k([0,T]\times\omega\times\mathbb{T})}. \tag{2.4.9}$$

Our method is by duality and requires nonsmooth profiles. Estimate (2.4.8) implies that the family J^ε_k can be extended *asymptotically* as a map from $\mathcal{L}^q_k$ to $L^q([0,+\infty[\times\mathbb{R}^d)$, so that the substitution $\theta = \psi(y)$ in $\mathcal{A}(t,y,\theta)$, which has no direct meaning for $\mathcal{A} \in \mathcal{L}^q_k$, makes sense *asymptotically*. More precisely, one has the following result.

PROPOSITION 2.4.3. *For each $\mathcal{U} \in \mathcal{L}^q_k$ there exists a bounded family u^ε in $L^q([0,+\infty[\times\mathbb{R}^d)$ such that for each $\delta > 0$, there is an $\varepsilon_\delta > 0$ and a trigonometric polynomial $\mathcal{U}_\delta$ which vanishes near $\mathcal{S}_k$, such that*

$$\|\mathcal{U} - \mathcal{U}_\delta\|_{\mathcal{L}^q_k} \;\leq\; \delta\,, \tag{2.4.10}$$

and

$$\forall \varepsilon \in]0,\varepsilon_\delta]\,,\quad \|u^\varepsilon - J^\varepsilon_k(\mathcal{U}_\delta)\|_{L^q([0,+\infty[\times\mathbb{R}^d)} \;\leq\; \delta\,. \tag{2.4.11}$$

When u^ε satisfies (2.4.10) (2.4.11), we write $u^\varepsilon \sim \tilde{J}^\varepsilon_k(\mathcal{U})$ in L^q. If $u^\varepsilon \sim \tilde{J}^\varepsilon_k(\mathcal{U})$ and $v^\varepsilon \sim \tilde{J}^\varepsilon_k(\mathcal{U})$ in L^q, then $u^\varepsilon - v^\varepsilon$ converges strongly to 0 in $L^q([0,+\infty[\times\mathbb{R}^d)$. More generally, we write $u^\varepsilon \sim v^\varepsilon$ in L^q, when u^ε and v^ε are bounded families in $L^q([0,+\infty[\times\mathbb{R}^d)$ such that $u^\varepsilon - v^\varepsilon$ converges strongly to 0 in $L^q([0,+\infty[\times\mathbb{R}^d)$.

The notations are consistent. When $\mathcal{U} \in \mathcal{L}^{s,q}_k$ and $s > 1/q$, $J^\varepsilon_k(\mathcal{U})$ is defined for each ε, and

$$J^\varepsilon_k(\mathcal{U}) \;\sim\; \tilde{J}^\varepsilon_k(\mathcal{U}) \quad \text{in} \quad L^q([0,+\infty[\times\mathbb{R}^d) \tag{2.4.12}$$

in the sense defined above.

There is a similar treatment for functions on $\mathbb{R}^d$, i.e. for initial data. Given $\mathcal{U}_0 \in L^2(\omega\times\mathbb{T})$, one defines bounded families in $L^2(\mathbb{R}^d)$, denoted $u^\varepsilon_0 \sim \tilde{J}^\varepsilon_0(\mathcal{U}_0)$, and such that

$$J^\varepsilon_0(\mathcal{U}_0)\,(y) \;:=\; \mathcal{U}_0(y,\psi(y)/\varepsilon)\,, \tag{2.4.13}$$

when $\mathcal{U}_0$ is smooth, or more generally whenever the substitution $\theta = \psi(y)$ makes sense, for example when $\mathcal{U}_0 \in L^2(\omega\times\mathbb{T})$ and $\partial_\theta\mathcal{U}_0 \in L^2(\omega\times\mathbb{T})$.

2.5. Sketch of the main results

Consider Cauchy data such that

$$u_0^\varepsilon \sim \underline{u}_0 + \tilde{J}_0^\varepsilon(\mathcal{U}_0) \quad \text{in } L^2, \tag{2.5.1}$$

where $\underline{u}_0 \in L^2(\mathbb{R}^d)$, $\mathcal{U}_0 \in L^2(\omega \times \mathbb{T})$ and the mean value $\int_{\mathbb{T}} \mathcal{U}_0(y, \theta)\, d\theta$ vanishes for almost all $y \in \omega$.

The family u_0^ε is bounded in $L^2(\mathbb{R}^d)$, and therefore, there is a unique bounded family

$$u^\varepsilon \in C^0([0, \infty[\,; L^2(\mathbb{R}^d)) \cap L^{p+1}([0, \infty[\times\mathbb{R}^d)$$

such that u^ε satisfies (2.1.1) with the initial condition

$$u^\varepsilon\big|_{t=0} = u_0^\varepsilon\,. \tag{2.5.2}$$

An abbreviated version of the main result is the following. Suppose that the assumptions in §§2.1, 2.2, 2.3 are satisfied.

THEOREM 2.5.1. *There are unique* $\mathcal{U}_k \in \mathcal{L}_k^{p+1} \cap C^0([0, \infty[\,; L^2(\omega \times \mathbb{T}))$ *and* $\underline{u} \in C^0([0, +\infty[; L^2(\mathbb{R}^d)) \cap L^{p+1}([0, \infty[\times\mathbb{R}^d)$, *such that for all* $T < +\infty$,

$$u^\varepsilon \sim \underline{u} + \sum_{k=1}^{k_0} \tilde{J}_k^\varepsilon(\mathcal{U}_k) \quad \text{in} \quad L^{p+1}([0, T] \times \mathbb{R}^d)\,. \tag{2.5.3}$$

When $\partial_\theta \mathcal{U}_0 \in L^2(\omega \times \mathbb{T})$, *the profiles* $\mathcal{U}_k$ *belong to* $\mathcal{L}_k^{s,p+1}$ *for all* $s < 2/(p+1)$ *and one has the stronger result*

$$u^\varepsilon \sim \underline{u} + \sum_{k=1}^{k_0} J_k^\varepsilon(\mathcal{U}_k) \quad \text{in} \quad L^{p+1}([0, T[\times\mathbb{R}^d)\,. \tag{2.5.4}$$

The first step of the analysis is to introduce the weak limits $\underline{u}$ in L^{p+1} and $\underline{f}$ in $L^{1+1/p}$ of u^ε and $f^\varepsilon := f(u^\varepsilon)$, extracting subsequences if necessary. Extracting further subsequences, we introduce profiles $\mathcal{U}_k$ and $\mathcal{F}_k$ such that for all $\mathcal{A} \in \mathcal{P}$ which vanish on a neighborhood of $\mathcal{S}_k$

$$\int u^\varepsilon\, \overline{J_k^\varepsilon(\mathcal{A})}\, dt\, dx \longrightarrow \int \mathcal{U}_k\, \overline{\mathcal{A}}\, dt\, dx\, d\theta\,, \tag{2.5.5}$$

and

$$\int f^\varepsilon\, \overline{J_k^\varepsilon(\mathcal{A})}\, dt\, dx \longrightarrow \int \mathcal{F}_k\, \overline{\mathcal{A}}\, dt\, dx\, d\theta\,. \tag{2.5.6}$$

By continuity, the convergence (2.5.5) (resp. (2.5.6)) extends to $\mathcal{A} \in \mathcal{L}_k^{1+1/p}$ (resp. $\mathcal{A} \in \mathcal{L}_k^{p+1}$).

Alternatively one can replace J_k^ε by I_k^ε in (2.5.5), (2.5.6) to find

$$\int f^\varepsilon\, \overline{I_k^\varepsilon(\mathcal{A})}\, dt\, dx \longrightarrow \int \mathcal{F}_k\, \overline{\mathcal{A}}\, dt\, dx\, d\theta\,. \tag{2.5.7}$$

It is crucial to observe that (2.5.7) holds for $\mathcal{A} \in \mathcal{P}$ which vanishes on a neighborhood of supercritical points $\mathcal{S}_k^+(p+1)$. This is true because f^ε is bounded in $L^{1+1/p}$ and $I_k^\varepsilon(\mathcal{A})$ is bounded in L^{p+1} for such $\mathcal{A}$, thanks to Theorem 2.3.5.

PROPOSITION 2.5.2. *The profiles $\mathcal{U}_k \in \mathcal{L}_k^{p+1}$ satisfy the polarization condition*

$$P_k\,\mathcal{U}_k = \mathcal{U}_k\,. \tag{2.5.8}$$

Moreover, $\mathcal{F}_k \in \mathcal{L}_k^{1+1/p}$ and $\mathcal{U}_k$ satisfies on $G\backslash\mathcal{S}_k^+(p+1)\times\mathbb{T}$

$$\partial_t\mathcal{U}_k \;+\; P_k\,\mathcal{F}_k \;=\; 0\,, \qquad \mathcal{U}_k\big|_{t=0} \;=\; P_k\mathcal{U}_0\,. \tag{2.5.9}$$

The weak limits satisfy $\underline{u} \in L^{p+1}([0,\infty[\times\mathbb{R}^d)$, $\underline{f} \in L^{1+1/p}([0,\infty[\times\mathbb{R}^d)$ and

$$L\underline{u} \;+\; \underline{f} \;=\; 0 \quad on\ [0,\infty[\times\mathbb{R}^d\,, \qquad \underline{u}\big|_{t=0} \;=\; \underline{u}_0\,. \tag{2.5.10}$$

In (2.5.8) (2.5.9), $P_k(y)$ denotes the spectral projector of $A(d\psi(y))$ corresponding to the eigenvalue $\lambda_k(d\psi(y))$. Formula (2.5.9) follows from (2.5.7) and the known asymptotic behavior of $L(I_k^\varepsilon(\mathcal{A}))$ (see §6). The restriction to the complement of $\mathcal{S}_k^+(p+1)$ comes from the fact that the test profiles in (2.5.7) are required to vanish on $\mathcal{S}_k^+(p+1)$.

The proof of (2.5.3) uses the energy identity

$$\begin{aligned} \| u^\varepsilon(t) - v^\varepsilon(t) \|_{L^2}^2 \;-\; 2\,\mathrm{Re}\int_{[0,t]\times\mathbb{R}^d} L(v^\varepsilon - u^\varepsilon)\cdot(\overline{v^\varepsilon} - \overline{u^\varepsilon}) \;=& \\ \| u^\varepsilon(0) - v^\varepsilon(0) \|_{L^2}^2\,.& \end{aligned} \tag{2.5.11}$$

This implies

$$\begin{aligned} &\| u^\varepsilon(t) - v^\varepsilon(t) \|_{L^2}^2 \;+\; c\,\| u^\varepsilon - v^\varepsilon \|_{L^{1+p}([0,t]\times\mathbb{R}^d)}^{1+p} \;\le \\ &\| u^\varepsilon(0) - v^\varepsilon(0) \|_{L^2}^2 \;+\; 2\,\mathrm{Re}\int_{[0,t]\times\mathbb{R}^d} (Lv^\varepsilon + f(v^\varepsilon))\cdot(\overline{v^\varepsilon} - \overline{u^\varepsilon})\;dt\,dx\,. \end{aligned} \tag{2.5.12}$$

We chose $v^\varepsilon \sim \underline{u} + \sum \tilde{J}_k^\varepsilon(\mathcal{U}_k)$. Note that the profiles are not yet known to be smooth so we need the extended definition $\tilde{J}$. An important step is to show, using Proposition 2.5.2, that one can also choose v^ε so that

$$Lv^\varepsilon \sim \underline{f} + \sum \tilde{J}_k^\varepsilon(P_k\mathcal{F}_k) \;\text{ in }\; L^{1+1/p}. \tag{2.5.13}$$

The next important step is to analyse $f(v^\varepsilon)$. For $v^\varepsilon \sim \underline{u} + \sum \tilde{J}_k^\varepsilon(\mathcal{U}_k)$, one has

$$f(v^\varepsilon) \;\sim\; \underline{\mathcal{E}}(\underline{v},\mathcal{U}_*) \;+\; \sum_k \tilde{J}_k^\varepsilon(P_k\mathcal{E}_k(\underline{v},\mathcal{U}_*)) \;+\; h^\varepsilon\,, \tag{2.5.14}$$

where $\mathcal{U}_* := (\mathcal{U}_1,\dots,\mathcal{U}_{k_0})$, $\underline{\mathcal{E}}$, and $\mathcal{E}_k$ are nonlinear operators mapping $L^{p+1}\times\prod\mathcal{L}_l^{p+1}$ to $L^{1+1/p}$,$L^{1+1/p}$, and $\mathcal{L}_k^{1+1/p}$ respectively. The family h^ε is bounded in $L^{1+1/p}$ and has no oscillations in the characteristic variety of L. This follows from the nonresonance Assumption 2.2.5. Since $L(u^\varepsilon - v^\varepsilon)$ is bounded in $L^{1+1/p}$ by (2.5.13), Proposition 5.2.2 and Corollary 5.2.4 show that

$$\lim_{\varepsilon\to 0}\int h^\varepsilon(u^\varepsilon - v^\varepsilon) = 0\,.$$

Thus, the weak limit and the profiles of $u^\varepsilon - v^\varepsilon$ vanish. This implies that

$$\int_{[0,t]\times\mathbb{R}^d} \Big(\underline{f} + \underline{\mathcal{E}}(\underline{u},\mathcal{U}_*) + \sum \tilde{J}_k^\varepsilon(P_k(\mathcal{F}_k + \mathcal{E}_k(\underline{u},\mathcal{U}_*))\Big)\cdot(\overline{v^\varepsilon}-\overline{u^\varepsilon})\,dt\,dx \longrightarrow 0\,. \tag{2.5.15}$$

Thus the right hand side of (2.5.12) tends to zero and $u^\varepsilon - v^\varepsilon$ tends to zero strongly in L^{1+p} ending the proof of (2.5.3).

The strong convergence implies that $f(u^\varepsilon) \sim f(v^\varepsilon)$ in $L^{1+1/p}$. Therefore (2.5.13) yields

$$\underline{f} = \underline{\mathcal{E}}(\underline{u},\mathcal{U}_*)\,, \qquad P_k\mathcal{F}_k = P_k\,\mathcal{E}_k(\underline{u},\mathcal{U}_*)\,. \tag{2.5.16}$$

Proposition 2.5.2 then yields the following equations for the profiles

$$\begin{aligned} L\underline{u} + \underline{\mathcal{E}}(\underline{u},\mathcal{U}_*) &= 0\,,\\ P_k\,\mathcal{U}_k &= \mathcal{U}_k,\\ \partial_t\mathcal{U}_k + P_k\,\mathcal{E}_k(\underline{u},\mathcal{U}_*) &= 0 \quad \text{on} \quad G\backslash\mathcal{S}_k^+(p+1)\times\mathbb{T}\,. \end{aligned} \tag{2.5.17}$$

The initial conditions are

$$\underline{u}\big|_{t=0} = \underline{u}_0\,, \quad \mathcal{U}_k\big|_{t=0} = P_k\,\mathcal{U}_0\,. \tag{2.5.18}$$

Knowing that $\mathcal{U}_*$ satisfies the profile equations (2.5.17) (2.5.18), one shows that $\mathcal{U}_k \in \mathcal{L}_k^{s,p+1}$ for all $s < 2/(p+1)$ when $\partial_\theta\mathcal{U}_0 \in L^2$. In that case, the $J_k^\varepsilon(\mathcal{U}_k)$ are defined for each ε, and (2.4.12) implies the stronger form (2.5.4) of the asymptotics.

Note that $\mathcal{E}_k(\underline{u},\mathcal{U}_*)$ belongs to $\mathcal{L}_k^{1+1/p}$. Near supercritical points, $\mathcal{L}_k^{1+1/p}$ is *not* included in L^1_{loc}, since this would imply that $L^\infty \subset \mathcal{L}_k^{1+p}$ near $\mathcal{S}_k^+(1+p)$ which is impossible from Definition 2.4.1. Thus the meaning of the third equation in (2.5.17) is not clear near $\mathcal{S}_k^+(p+1)$. However, the third equation *does* determine $\mathcal{U}_k$ because near supercritical points, *strong* dissipation intervenes, and there, the *absorption* phenomenon occurs. This suggests the following terminology.

DEFINITION 2.5.3. *For $y \in \omega$, the k-time of absorption along the ray from y is the smallest value of $t > 0$ such that $(t,y) \in \mathcal{S}_k^+(p+1)$ the set of $p+1$-supercritical point in $\mathcal{S}_k$, i.e. satisfying $\mu_k(t,y) \geq 2/(p-1)$. It is denoted by $T_k^{abs}(y)$.*

Then, $1/T_k^{abs}(y)$ is the largest positive eigenvalue of $g_k'(y)$ of multiplicity larger than or equal to $2/(p-1)$. $T_k^{abs}(y)$ is infinite if there is no such eigenvalue.

Introduce

$$\begin{cases} G_k^- := \{\,(t,y) \in]0,+\infty[\times\omega_0 \mid t < T_k^{abs}(y)\,\}\\[1ex] G_k^+ := \{\,(t,y) \in]0,+\infty[\times\omega_0 \mid t > T_k^{abs}(y)\,\} \end{cases} \tag{2.5.19}$$

T_k^{abs} is C^∞ on ω_0 where the eigenvalues of $g_k'(y)$ are smooth. Thus G_k^- is open. Moreover, $G\backslash\mathcal{S}_k^+(p+1)$ is the union of G_k^-, G_k^+ and a negligible set.

THEOREM 2.5.4. *With notations as in Theorem 2.5.1, the profiles $\mathcal{U}_k$ vanish on $G_k^+\times\mathbb{T}$, and the third equation in (2.5.17) holds in L^1_{loc} on the domains $G_k^-\times\mathbb{T}$.*

2.6. Remarks and Examples

1) Note that $g_k'(y) = \lambda_k''(d\psi(y))\,\psi''(y)$. Since λ_k' is homogeneous of degree 0, $\lambda_k''(\xi)$ has always a nontrivial kernel which contains ξ. Thus 0 is always an eigenvalue of $g_k'(y)$. As a consequence, the multiplicities μ_k satisfy

$$(2.6.1) \qquad \forall k \in \{1, \dots, k_0\}, \forall (t,y) \in \mathcal{S}_k, \quad 1 \le \mu_k(t,y) \le d-1 .$$

In space dimension $d = 2$, the only possibility is $\mu_k = 1$ and Assumption 2.3.3 is always satisfied.

2) The Lagrangian Λ_k associated to the eiconal equation (2.2.2) is the set of (t, x, τ, ξ) such that

$$(2.6.2) \qquad x = q_k(t,y), \quad \xi = d\psi(y), \quad \tau = \lambda_k(d\psi(y)) .$$

It has the defining phase function

$$(2.6.3) \qquad \Phi_k(t,x,y,\xi) := t\,\lambda_k(\xi) + (x-y)\cdot\xi + \psi(y) .$$

$\mathcal{S}_k$ corresponds to the set of points in Λ_k where the projection $\pi : (t,x,\tau,\xi) \mapsto (t,x)$ restricted to Λ_k, is not of maximal rank $d+1$. The multiplicity μ_k is equal to the order of vanishing of $\det d(\pi|_{\Lambda_k})$. It is larger than or equal to the dimension of the kernel of $d(\pi|_{\Lambda_k})$. In §3.5 we show that for the wave equation, these two numbers are equal.

3) If $p \ge 3$, all points in $\mathcal{S}_k$ are supercritical, since then $2/(p-1) \le 1 \le \mu_k(t,y)$. Thus $T_k^{abs}(y)$ is the first time of focusing along the ray.

In the extreme opposite case, all points in $\mathcal{S}_k$ are subcritical when $p - 1 < +2/(d-1)$, since then $2/(p-1) > d-1 \ge \mu_k(t,y)$, see (2.6.1). In this case T_k^{abs} is infinite.

4) For radial focusing in a wave equation, $\lambda(\xi) := |\xi|$ and $\psi(y) := \chi(|y|)$. Then, g' has only one positive eigenvalue equal to $1/|y|$, of multiplicity $d-1$. Thus absorption occurs if and only if $p \ge 1 + 2/(d-1)$. The if part was proved in [JMR 2].

5) In many examples, $\mu_k(t,y) = 1$ for all points $(t,y) \in \mathcal{S}_k$. Then Assumption 2.3.3 is satisfied. This happens when all the rays meet the caustic at generic points (folds) (see Remark 3.5.2). It also occurs when $\pi|_{\Lambda_k}$ has a cusp. In these two cases The critical exponent is $p = 3$. If $p < 3$, $T_k^{abs}(y)$ is infinite. Conversely if $p \ge 3$, absorption occurs at the first encounter with caustics. Examples in space dimension 2 for the wave equation are the familiar phases

$$(2.6.4) \quad \psi_1(y_1,y_2) = y_2 + y_1^2, \quad \psi_2(y_1,y_2) = a_1 y_1^2 + a_2 y_2^2 \quad \text{with} \quad 0 < a_1 < a_2 .$$

For ψ_1 the caustic set is the manifold $\mathcal{C}$ with equation $x_1^2 = (t^{2/3} - 1)^3$, which has a cusp singularity at $x_1 = 0, t = 1$, corresponding to the rays from $y_1 = 0$.

6) Again for the wave equation, $\lambda_k(\xi) = |\xi|$, in $\mathbb{R}^d$ consider

$$(2.6.5) \qquad \psi_3(y) := y_d + (y_1^2 + \cdots + y_{d-1}^2)/2 =: y_d + |y'|^2/2.$$

For $y' \neq 0$, there are two distinct positive eigenvalues : $1/\delta$ of multiplicity $d-2$ and $1/\delta^3$ of multiplicity 1, where $\delta^2 := 1 + |y'|^2$. For $y' = 0$, these eigenvalues are equal. Therefore, $(t = 1, y' = 0)$ is a crossing point for two different eigenvalues c's, but it is the only point where multiplicities are not locally constant. Nevertheless Assumptions 2.2.2 and 2.3.3 are still satisfied.

There are two focusing times, δ and δ^3. Accordingly, the caustic set has two components. The first one is the plane $\mathcal{C}_1 = \{x' = 0\}$. The second is the manifold $\mathcal{C}_2$ with equation $x'^2 = (t^{2/3} - 1)^3$, which has a singular locus at $x' = 0, t = 1$, corresponding to the rays from $y' = 0$. Since $\delta < \delta^3$, the rays cross $\mathcal{C}_1$ before $\mathcal{C}_2$.

If $p \geq 1 + 2/(d-2)$, then $\mathcal{C}_1$ is supercritical. Oscillations are absorbed at $\mathcal{C}_1$ and never reach $\mathcal{C}_2$. If $p < 1 + 2/(d-2)$, then both $\mathcal{C}_1$ and $\mathcal{C}_2$ are subcritical. Oscillations cross $\mathcal{C}_1$ and $\mathcal{C}_2$.

3. L^p estimates for oscillatory integrals

The aim of this section is to provide uniform L^q estimates for oscillatory integrals such as $I^\varepsilon(\mathcal{A})$. Because the question is interesting in its own, we consider more general classes of oscillatory integrals. At the end of this section the results are applied to the context developed in §2. The estimates are proved under general assumptions which are satisfied for the examples of generic caustics, cusps, and spherical focusing.

3.1 Statement of the main estimate

Consider integrals of the form

$$(3.1.1) \qquad u^\varepsilon(z) := \varepsilon^{-n/2} \int_{\mathbb{R}^n} e^{i\,\Phi(z,\alpha)/\varepsilon}\, a(z,\alpha,\varepsilon)\, d\alpha \quad , z \in \mathbb{R}^m,$$

where Φ and a are C^∞ functions on $\mathbb{R}^m \times \mathbb{R}^n$ and $\mathbb{R}^m \times \mathbb{R}^n \times [0,1]$ respectively. Φ is real valued and a is compactly supported. Our goal is to obtain bounds for the L^q-norm of u^ε.

Results in this direction can be found in [R 1], [R 2] and [Sv]. They concern phases of the form $\Phi(z,\alpha) = z \cdot \alpha - h(\alpha)$ with h real analytic and convex, or $n = m = 1$ and h'' having only isolated finite order zeros. For example, it is proved in [R 2] that, for amplitudes with small support,

$$(3.1.2) \qquad \sup_{\varepsilon \in]0,1]} |u^\varepsilon(z))| \leq C\, \big|\det h''\big(\alpha(z)\big)\big|^{-1/2}$$

where, $\alpha(z)$ is the unique critical point of $\alpha \mapsto \Phi(z,\alpha)$. Note that the stationary phase theorem implies that a control of u^ε by $|\det h''|^{-1/2}$ is the best estimate that one can expect. However, an estimate like (3.1.2) is not true in general. In the simplest case where $n = m = 1$ and $\Phi = z\alpha - \alpha^3$ which leads to Airy integrals, there is no critical point in the shadow region $z < 0$ but u^ε is not uniformly bounded in this region. We refer to [R 1] for a substitute for (3.1.2) for phases of the form $z\alpha - \alpha^k$.

The convexity assumptions are not always satisfied. For example, a typical phase for which the caustic set is a cusp, is $\Phi(z_1, z_2, \alpha) = z_1\alpha + z_2\alpha^2 + \alpha^4$. For this and the phases (2.3.3), for the convexity assumptions are not satisfies and we are not able to prove estimates like (3.1.2). However, we can determine the supremum of the real numbers q such that the family u^ε is bounded in L^q, and that is sufficient for our application to oscillations for dissipative wave equations.

Using a partition of unity, it is sufficient to assume that the amplitude a in (3.1.1) is supported in a small neighborhood of a point $\underline{\rho} := (\underline{z}, \underline{\alpha})$. Because $u^\varepsilon =$

$O(\varepsilon^\infty)$ when the phase has no stationary points in a neighborhood of the support of a, and $u^\varepsilon = O(1)$ for a nondegenerate stationary point, we study the remaining case where $\underline{\rho}$ is a *degenerate stationary point*, i.e.

$$d_\alpha \Phi(\underline{\rho}) = 0\,, \qquad 0 = \det \Phi''_{\alpha\alpha}(\underline{\rho}) := D(\underline{\rho})\,, \tag{3.1.3}$$

where $\Phi''_{\alpha\alpha}$ is the the Hessian matrix

$$\Phi''_{\alpha\alpha}(z,\alpha) := \left[\frac{\partial^2 \Phi}{\partial\alpha_j\,\partial\alpha_k}(z,\alpha)\right]_{1\le j,\,k\le n} \tag{3.1.4}$$

The behavior of u^ε depends on the structure of the set of degenerate stationary points. We consider simple cases which naturally occur in the study of the integrals $I^\varepsilon(\mathcal{A})$ from §2. The first assumption is that the phase Φ is nondegenerate (see e.g. [Hö 1]), which means that Φ'_α is of maximal rank at $\underline{\rho}$,

$$\operatorname{rank}\left(d_{(z,\alpha)}\,\Phi'_\alpha(\underline{z},\underline{\alpha})\right) = n\,. \tag{3.1.5}$$

Thus, near $\underline{\rho}$, the set of stationary points

$$C_\Phi := \{\,(z,\alpha)\;:\;\Phi'_\alpha(z,\alpha) = 0\,\} \tag{3.1.6}$$

is a smooth manifold of dimension m. The set of degenerate stationary points is

$$S_\Phi := \{\,(z,\alpha)\in C_\Phi\;:\;D(z,\alpha) = 0\,\}\,. \tag{3.1.7}$$

First we give a necessary condition for the family u^ε to be bounded in L^q.

THEOREM 3.1.1. *There is a neighborhood ω of $\underline{\rho}$ so that if an amplitude $a \in C_0^\infty(\omega\times[0,1])$ defines a family* (3.1.1) *that is bounded in L^q with $q > 2$, then the restriction of $|D|^{1-q/2}(\rho)\,|a(\rho,0)|^q$ to C_Φ is integrable.*

Next we turn to sufficient conditions. At the extreme points of our parameter domain we use the Lorentz spaces. Recall that a measurable function u belongs to the weak-$L^p := L^{p,\infty}$ when

$$\|u\|_{p,\infty} := \Big(\sup_{s>0} s^p \operatorname{meas}\{x, |u(x)| > s\}\Big)^{1/p} < +\infty,$$

and that locally $L^q_{loc} \subset L^{p,\infty}_{loc}$ for every $q \ge p$.

Denote by D_* the restriction of D to C_Φ.

THEOREM 3.1.2. *Suppose that there exist a neighborhood $\mathcal{O}$ of $\underline{z}$ and an integer ℓ such that for almost all $z \in \mathcal{O}$ the number of critical points $(z,\alpha)\in C_\Phi$ is not larger than ℓ. Suppose in addition that the dimension of* $\ker \Phi''_{\alpha\alpha}(\underline{\rho})$ *is equal to* 1.

If $r > 0$ is such that $D_*^{-r} \in L^{1,\infty}$ [resp. L^1] on a neighborhood of $\underline{\rho}$ in C_Φ, then there exists a neighborhood ω of $\underline{\rho}$ such that for all $a \in C_0^\infty(\omega\times[0,1])$ the family u^ε is bounded in $L^{2+2r,\infty}$ [resp. L^{2+2r}].

EXAMPLE. When the phase Φ is real analytic, the first assumption in Theorem 3.1.2 is automatically satisfied. Thus when the dimension of $\ker \Phi''_{\alpha\alpha}(\underline{\rho})$ is equal to one and Φ is real analytic, Theorems 3.1.1 and 3.1.2 give a a complete answer to our question.

When the dimension of $\ker \Phi''_{\alpha\alpha}(\underline{\rho})$ is larger than one, our results are less complete. The following theorem covers the important case of radial focusing in $\mathbb{R}^d$, $d \geq 3$.

THEOREM 3.1.3. *Suppose that $\mathcal{S}_\Phi$ is a smooth submanifod of codimension one in C_Φ, the dimension $\nu := \dim \ker \Phi''_{\alpha\alpha}(\rho)$ is a constant at least equal to 2 for ρ in a neighborhood of $\underline{\rho}$ in $\mathcal{S}_\Phi$, and D_* vanishes exactly to order $\mu = \nu$ on $\mathcal{S}_\Phi$.*

Then there exists a neighborhood ω of $\underline{\rho}$ such that for all $a \in C_0^\infty(\omega \times [0,1])$ the family u^ε is bounded in $L^{2+2/\mu,\infty}$.

REMARK 3.1.4. When $\mathcal{S}_\Phi$ is a smooth submanifold of codimension one in C_Φ and D_* vanishes exactly to order μ on $\mathcal{S}_\Phi$, then $D_*^{-2/\mu} \in L^{1,\infty}$. Thus when the dimension of $\dim \ker \Phi''_{\alpha\alpha}(\underline{\rho}) = 1$ and the number of critical points is bounded, or when this dimension is locally constant and equal to μ, the family u^ε is bounded in $L^{2+2/\mu,\infty}$, provided that a is supported in a small neighborhood of $\underline{\rho}$. In particular u^ε is bounded in L^q for all $q < 2 + 2/\mu$. The index $q_c = 2 + 2/\mu$ is sharp, since Theorem 3.1.1 implies that whem $a(\underline{\rho}, 0) \neq 0$, the family u^ε is not bounded in L^q for $q \geq 2 + 2/\mu$.

REMARK 3.1.5. The assumptions are invariant under change to equivalent phase functions and have the following intrinsic formulation. Assumption (3.1.5) implies that

$$\Lambda := \iota(C_\Phi) \subset T^*\mathbb{R}^m \,, \qquad \text{where} \qquad \iota : (z,\alpha) \mapsto (z, \Phi'_z(z,\alpha)) \tag{3.1.8}$$

is a smooth Lagrangian manifold and that ι is a local diffeomorphism from C_Φ to Λ. Let π denote the restriction to Λ of the projection $(z,\zeta) \mapsto z$. The kernel of π' at $(z,\zeta) = \iota(\rho)$ is exactly the image by $\iota'(\rho)$ of the space $\{0\} \times \ker \Phi''_{\alpha\alpha}$ which is tangent to C_Φ at ρ. In particular,

$$\begin{cases} \iota(\mathcal{S}_\Phi) = \mathcal{S} := \{(z,\zeta) \ ; \ \operatorname{rank} \pi'(z,\zeta) < m\}\,, \\ \dim \ker \pi'(z,\zeta) \ = \ \dim \ker \Phi''_{\alpha\alpha}(\rho)\,, \quad \text{for} \quad (z,\zeta) = \iota(\rho) \in \mathcal{S}\,. \end{cases} \tag{3.1.9}$$

Therefore, the assumptions in Theorem 3.1.2 are : 1) for almost all z in a neighborhood of $\underline{z}$ the number of preimages of x by π in Λ is less than or equal to ℓ, 2) $\ker \pi'(\underline{z}, \underline{\zeta})$ has dimension 1. These assumptions must be satisfied by the germ of C_Φ near $\underline{\rho}$. The proves invariance under change of phase function which define the same germ of Lagrangian Λ.

Similarly, the assumptions in Theorem 3.1.3 are 1) $\mathcal{S}$ is a smooth submanifold of codimension 1 in Λ, 2) $\ker \pi'(z,\zeta)$ has constant dimension ν for (z,ζ) in a neighborhood of $(\underline{z}, \underline{\zeta})$ in $\mathcal{S}$, 3) $\det \pi'$ vanishes exactly to order ν on $\mathcal{S}$. This shows invariance under change of phase functions.

3.2. Proof of Theorem 3.1.1

Using the Theorem of equivalence of phase functions (see [Hö 1] [Du]), shrinking neighborhoods and multiplying u^ε by $e^{i\varphi(z)/\varepsilon}$ if necessary, we can assume that $m = n$ and on a neighborhood $\omega = \Omega \times U$ of $\underline{\rho} = (\underline{z}, \underline{\zeta})$, and Φ is of the form

$$\Phi(z,\alpha) \; = \; x \cdot \alpha \; - \; h(\alpha)\,, \tag{3.2.1}$$

with h smooth on a neighborhood of $\overline{U}$. In this case, $C_\Phi = \{(z,\alpha) \in \Omega \times U \mid z = h'(\alpha)\}$, and $D_*(\alpha) = \det h''(\alpha)$. Consider $a \in C_0^\infty(\omega \times [0,1])$.

For $f \in C_0^\infty(\Omega)$, one has

$$\int f(z)\, |u^\varepsilon(z)|^2\, dz \; = \; \int F^\varepsilon(\eta, \alpha, \alpha + \varepsilon\eta)\; e^{i\,(h(\alpha) - h(\alpha+\varepsilon\eta))/\varepsilon}\, d\xi\, d\eta\,, \tag{3.2.2}$$

where

$$F^\varepsilon(\eta,\alpha,\alpha') \; := \; \int e^{i\,z\eta} f(z)\, a(z,\alpha,\varepsilon)\, \overline{a(z,\alpha',\varepsilon)}\, dz \; = \; O((1+|\eta|)^{-\infty})\,. \tag{3.2.3}$$

The estimate is uniform, and Lebesgues' dominated convergence theorem implies that

$$\lim_{\varepsilon \to 0} \int f(z)\, |u^\varepsilon(z)|^2\, dx \; = \; (2\pi)^d \int f(h'(\alpha))\, |a(h'(\alpha),\alpha,0)|^2\; d\alpha\,. \tag{3.2.4}$$

Therefore, if the family u^ε is bounded in L^q, there is a constant C such that for all $f \in C_0^\infty(\Omega)$,

$$\Big| \int f(h'(\xi))\, |a(h'(\xi),\xi)|^2\, d\xi \Big| \; \le \; C\, \| f \|_{L^r(\Omega)}\,, \qquad r := q/(q-2)\,. \tag{3.2.5}$$

Introduce the measure λ on Ω which is the image of $|a(h'(\alpha),\alpha,0)|^2\, d\alpha$ by h'. The left hand side of (3.2.5) is equal to $\big| \int f(z)\, \lambda(dz) \big|$. Thus (3.2.5) implies that

$$\lambda \; = \; \ell(x)\, dx\,, \qquad \text{where} \quad \ell \in L^{r'}(\Omega)\,, \quad r' = q/2\,. \tag{3.2.6}$$

Introduce $\mathcal{S} = \{\alpha \in \overline{U} \;:\; D_*(\alpha) = 0\}$, where $D_* := \det h''$. Define $\mathcal{C} := h'(\mathcal{S}) \cap \overline{\Omega_1}$. For all $z \in \Omega \backslash \mathcal{C}$, the set of $\alpha \in \overline{U}$ such that $h'(\alpha) = z$ is finite and there is a neighborhood G of z such that $(h')^{-1}(G)$ is the union of finitely many pairwise disjoint open sets V_j, such that h' is a diffeomorphism from V_j onto G. This implies that on $\Omega \backslash \mathcal{C}$

$$\ell(z) \; = \; \sum_{\alpha \in (h')^{-1}(z)} \frac{|a(z,\alpha)|^2}{|D_*(\alpha)|} \; \ge \; 0\,. \tag{3.2.7}$$

Sard's theorem implies that the Lebesgue measure of $\mathcal{C}$ is equal to 0. Therefore the density ℓ is completely determined by (3.2.7). Since $\ell \in L^{r'}$, one has

$$\begin{aligned} \int \ell(z)^{r'}\, dz \; &= \; \int \ell(z)^{r'-1}\, \lambda(dz) \\ &= \; \int \ell(h'(\alpha))^{r'-1}\, |a(h'(\alpha),\alpha,0)|^2\, d\alpha \; < \; +\infty\,. \end{aligned} \tag{3.2.8}$$

For $\alpha \notin (h')^{-1}(\mathcal{C})$, (3.2.7) shows that

$$\ell(h'(\alpha))^{r'-1}\, |a(h'(\alpha),\alpha,0)|^2 \; \ge \; |D_*(\alpha)|^{1-q/2}\, |a(h'(\alpha),\alpha,0)|^q\,. \tag{3.2.9}$$

In addition, (3.2.6) implies that $\lambda(\mathcal{C}) = 0$, hence $(h')^{-1}(\mathcal{C})$ is a null set for the measure $|a(h'(\alpha),\alpha,0)|^2\,d\alpha$. Therefore (3.2.8) and (3.2.9) imply that

$$\int |D_*(\alpha)|^{1-q/2}\,|a(h'(\alpha),\alpha,0)|^q\,d\alpha \;<\; +\infty\,, \tag{3.2.10}$$

which finishes the proof of Theorem 3.1.1.

3.3 Proof of Theorem 3.1.2

1. By a linear change of variables, one can write $\alpha = (\alpha',\alpha'') \in \mathbb{R}\times\mathbb{R}^{n-1}$, so that $\Phi''_{\alpha',\alpha'}(\underline{\rho}) = 0$ and $\Phi''_{\alpha''\alpha''}(\underline{\rho})$ is invertible. Thus, near $\underline{\rho}$, the phase $\alpha'' \to \Phi(z,\alpha',\alpha'')$ has a unique critical point, $\alpha''(z,\alpha')$ which is nondegenerate. Therefore, if a is supported in a sufficiently small neighborhood of $\underline{\rho}$, the stationary phase theorem implies that

$$\varepsilon^{-(n-1)/2}\int_{\mathbb{R}^{n-1}} e^{i\,\Phi(z,\alpha',\alpha'')/\varepsilon}\; a(z,\alpha',\alpha'',\varepsilon)\; d\alpha'' \;=\; b(z,\alpha',\varepsilon)\; e^{i\,\Psi(x,\alpha')/\varepsilon}\,, \tag{3.3.1}$$

where b is C^∞ on $U'\times[0,1]$ and $\Psi(z,\alpha') := \Phi(z,\alpha',\alpha''(z,\alpha'))$. Thus

$$u^\varepsilon(z) \;=\; \varepsilon^{-\nu/2}\int_{\mathbb{R}^\nu} e^{i\,\Psi(z,\alpha')/\varepsilon}\; b(z,\alpha',\varepsilon)\; d\alpha'\,. \tag{3.3.2}$$

Since Ψ and Φ define the same germ of Lagrangian manifold, Ψ satisfies the same assumptions as Φ. Therefore, it is sufficient to prove Theorem 3.1.2 when $n = 1$ which we assume from now on.

2. The phase Φ is nondegenerate and $\underline{\rho} \in \mathcal{S}_\Phi$. Therefore, there are coordinates $z = (t,y) \in \mathbb{R}^{m-1}\times\mathbb{R}$ such that $\partial^2_{y\,\alpha}\Phi(\underline{\rho}) \neq 0$. Using the Theorem of equivalence of phase functions (see [Hö 1] [Hö 2] [Du]) with t as parameters reduces to the case where

$$\Phi(z,\alpha) \;:=\; y\,\alpha \;-\; h(t,\alpha)\;, \qquad z = (t,y) \in \mathbb{R}^{d-1}\times\mathbb{R}\,. \tag{3.3.3}$$

From now on, we assume that Φ is given by (3.3.3) on $\omega = \Omega\times\mathcal{I}$ where $\Omega = \Omega_{\underline{t}}\times\Omega_{\underline{y}}$ is a relatively compact neighborhood of $\underline{z} = (\underline{t},\underline{y})$, $\mathcal{I}$ is an open interval which contains $\underline{\alpha}$, h is C^∞ on $\overline{\Omega_{\underline{t}}}\times\overline{\mathcal{I}}$ and $h'_\alpha(\Omega_{\underline{t}}\times\mathcal{I}) \subset \Omega_{\underline{y}}$. In this case, C_Φ is the graph $y = h'_\alpha(t,\alpha)$ with (t,α) as coordinates. Furthermore, $D_*(t,\alpha) = h''_{\alpha,\alpha}(t,\alpha)$ and $\mathcal{S}_\Phi$ is the set of (t,α) such that $D_*(t,\alpha) = 0$.

LEMMA 3.3.1. *There is a negligible set $\mathcal{N} \subset \Omega$ such that for all $x \in \Omega\backslash\mathcal{N}$, the equation $\Phi'_\alpha(z,\alpha) = \lambda$ has at most ℓ solutions α in $\mathcal{I}$ for almost all $\lambda \in \mathbb{R}$.*

Proof. Shrinking neighborhoods if necessary, the assumptions in Theorem 3.1.2 imply that there is a negligible set $\mathcal{N}' \subset \mathcal{O}$ such that for all $z \in \mathcal{O}\backslash\mathcal{N}'$, the equation $\Phi'_\alpha(z,\alpha) = y - h'_\alpha(t,\alpha) = 0$ has at most ℓ roots in $\mathcal{I}$. Therefore, there is a negligible set $\mathcal{N}_{\underline{t}} \subset \Omega_{\underline{t}}$ such that for all $t \in \Omega_{\underline{t}}\backslash\mathcal{N}_{\underline{t}}$, the set $\{y : (t,y) \in \mathcal{N}'\}$ is negligible, implying that for almost all $y \in \mathcal{O}_{\underline{y}}$, the equation $h'_\alpha(t,\alpha) = y$ has at most ℓ roots in $\mathcal{I}$. Since $h'_\alpha(\mathcal{O}_{\underline{t}}\times\mathcal{I}) \subset \Omega_{\underline{y}}$, the lemma follows with $\mathcal{N} = \mathcal{N}_{\underline{t}}\times\Omega_{\underline{y}}$.

From now on, we fix $a \in C_0^\infty(\mathcal{O}\times\mathcal{I}\times[0,1])$.

3. Consider $\delta > 0$ and $z \in \Omega \backslash \mathcal{N}$. Introduce

$$\mathcal{I}(z,\delta) := \{\alpha \in \mathcal{I} \,:\, |\Phi'_\alpha(z,\alpha)| > \delta\}\,, \qquad \mathcal{J}(z,\delta) := \{\alpha \in \mathcal{I} \,:\, |\Phi'_\alpha(z,\alpha)| < 2\delta\}\,.$$

Fix $\chi \in C^\infty(\mathbb{R})$ such that $\chi(\lambda) = 0$ for $|\lambda| \leq 1$ and $\chi(\lambda) = 1$ for $|\lambda| \geq 2$. In the integral (3.1.1), use the partition of unity $1 = \chi(\Phi'_\alpha/\delta) + (1 - \chi(\Phi'_\alpha/\delta))$. This splits u^ε into into two pieces. The second integral is carried by $\mathcal{J}(z,\delta)$ and is estimated by

$$\| a \|_{L^\infty} \operatorname{meas} \mathcal{J}(z,\delta)/\sqrt{\varepsilon}\,. \tag{3.3.4}$$

In the first integral, one integrates by parts to find

$$\begin{aligned} i\sqrt{\varepsilon} \int e^{i\,\Phi/\varepsilon}\, &\partial_\alpha(a\,\chi(\Phi'_\alpha/\delta)/\Phi'_\alpha)\, d\alpha \\ &= \; i\sqrt{\varepsilon} \int e^{i\,\Phi/\varepsilon}\, a\, \partial_\alpha(\chi(\Phi'_\alpha/\delta)/\Phi'_\alpha)\, d\alpha \; + O(\sqrt{\varepsilon}\, \| \partial_\alpha a \|_{L^\infty}/\delta)\,. \end{aligned}$$

Note that $\partial_\alpha\big(\chi(\varphi'_\alpha/\delta)/\varphi'_\alpha\big) = O(|\varphi''_{\alpha,\alpha}|/(\varphi'_\alpha)^2)$ is supported in $\mathcal{I}(z,\delta)$. Therefore, the first term is estimated by

$$C\,\sqrt{\varepsilon} \int_{\mathcal{I}(z,\delta)} \frac{|\Phi''_{\alpha,\alpha}|}{|\Phi'_\alpha|^2}\, d\alpha \; + \; C\sqrt{\varepsilon}/\delta \tag{3.3.5}$$

Next, we apply to the function $1/\Phi'_\alpha$ the following lemma.

LEMMA 3.3.2. *Suppose that f is a bounded and C^1 function from an open set $O \subset \mathbb{R}$ to $\mathbb{R}$. Suppose for almost all $\lambda \in \mathbb{R}$ the equation $f(\alpha) = \lambda$ has at most ℓ solutions. Then*

$$\int_O |\, f'(\alpha)\,|\, d\alpha \;\leq\; 2\,\ell\, \|\, f\, \|_{L^\infty(O)}\,. \tag{3.3.6}$$

Estimate (3.3.6) follows from a more general result sometimes called the area formula. If O is an open subset of $\mathbb{R}^d$ and $f \in C^1(O;\mathbb{R}^d)$ such that for almost all $\lambda \in \mathbb{R}^d$, the number $\ell(\lambda)$ of points $\alpha \in O$ such that $f(\alpha) = \lambda$ is finite. Then

$$\int_O |\, \det f'(\alpha)\,|\, d\alpha \;=\; \int_{f(O)} \ell(\lambda)\, d\lambda\,, \tag{3.3.7}$$

This formula is proved combining Sard's Theorem to eliminate the critical values with the change of variables formula near regular values (see e.g. [E G]).

Thanks to Lemma 3.3.1, for $z \in \Omega\backslash\mathcal{N}$ one can apply Lemma 3.3.2 to the function $\alpha \mapsto 1/\varphi'_\alpha$ to find the estimate

$$\int_{\mathcal{I}(z,\delta)} \frac{|\Phi''_{\alpha,\alpha}|}{|\Phi'_\alpha|^2}\, d\alpha \;\leq\; C/\delta\,.$$

With (3.3.4) et (3.3.5), this proves that there is a constant C such that for all $\varepsilon\, \in]0,1]$, all $\delta > 0$ and all $z \in \Omega\backslash\mathcal{N}$

$$|\, u^\varepsilon(z)\,| \;\leq\; C\,\Big(\frac{\sqrt{\varepsilon}}{\delta} + \frac{\operatorname{meas} \mathcal{J}(z,\delta)}{\sqrt{\varepsilon}}\Big)\,. \tag{3.3.8}$$

4. Next apply Lemma 3.3.2 to the function $\alpha \mapsto \Phi'_\alpha(z,\alpha)$ when $z \notin \mathcal{N}$ to show that

$$\int_{\mathcal{J}(z,\delta)} |\Phi''_{\alpha,\alpha}(z,\alpha)|\, d\alpha \;\le\; 2\,\ell\,\delta\,.$$

Therefore

$$\begin{aligned} &\text{meas}\{\alpha \in \mathcal{J}(z,\delta)\,:\,|\Phi''_{\alpha,\alpha}(z,\alpha)| \ge \delta^2/\varepsilon\,\} \\ &\qquad \le \varepsilon/\delta^2 \int_{\mathcal{J}(z,\delta)} |\Phi''_{\alpha,\alpha}(z,\alpha)|\, d\alpha \;\le\; 2\ell\varepsilon/\delta\,. \end{aligned} \tag{3.3.9}$$

Introduce

$$\mathcal{K}(z,\delta,\varepsilon) := \big\{\alpha \in \mathcal{I}\,:\,|\Phi'_\alpha(z,\alpha)| \le 2\delta \quad\text{and}\quad |\Phi''_{\alpha,\alpha}(z,\alpha)| \le \delta^2/\varepsilon\,\big\}\,. \tag{3.3.10}$$

Then, (3.3.8) and (3.3.9) imply that there is a constant C_1 such that for all $\varepsilon \in]0,1]$, all $\delta > 0$, and all $z \in \Omega\backslash\mathcal{N}$

$$|\,u^\varepsilon(x)\,| \;\le\; C_1\,\Big(\frac{\sqrt{\varepsilon}}{\delta} + \frac{\text{meas}\,\mathcal{K}(x,\delta,\varepsilon)}{\sqrt{\varepsilon}}\Big)\,. \tag{3.3.11}$$

5. For $s > 0$ and $\varepsilon \in]0,1]$, introduce

$$\mathcal{A}(s,\varepsilon) \;:=\; \big\{z \in \mathcal{O}\;:\;|\,u^\varepsilon(x)\,| \ge 2C_1 s\big\}\,. \tag{3.3.12}$$

Our goal is to estimate the measure of $\mathcal{A}(s,\varepsilon)$. Apply (3.3.11) with $\delta = \delta_s := \sqrt{\varepsilon}/s$ to find for $z \in \Omega\backslash\mathcal{N}$,

$$|\,u^\varepsilon(z)\,| \;\le\; C_1\;s \;+\; C_1\;\frac{\text{meas}\,\mathcal{K}(z,\sqrt{\varepsilon}/s,\varepsilon)}{\sqrt{\varepsilon}}\,.$$

Therefore, for $z \in \mathcal{A}(s,\varepsilon)\backslash\mathcal{N}$, one has

$$\text{meas}\,\mathcal{K}(x,\sqrt{\varepsilon}/s,\varepsilon) \;\ge\; \sqrt{\varepsilon}\;s \;:=\; \rho_s\,. \tag{3.3.13}$$

Introduce

$$\begin{aligned} &\mathcal{B}(s,\varepsilon) \;:= \\ &\quad\big\{\,(z,\alpha) \in \Omega \times \mathcal{I}\;:\;|\Phi'_\alpha(z,\alpha)| \le 2\,\sqrt{\varepsilon}/s \;\text{ and }\; |\Phi''_{\alpha\alpha}(z,\alpha)| \le 1/s^2\,\big\}\,. \end{aligned} \tag{3.3.14}$$

Then $\mathcal{B}(z,s,\varepsilon) := \{\alpha \in \mathcal{I}\,:\,(z,\alpha) \in \mathcal{B}(s,\varepsilon)\} \;=\; \mathcal{K}(z,\sqrt{\varepsilon}/s,\varepsilon)$. Since $\text{meas}\,\mathcal{N} = 0$, (3.3.13) implies that

$$\begin{aligned} \text{meas}\,\mathcal{B}(s,\varepsilon) \;&=\; \int \text{meas}\,\mathcal{B}(z,s,\varepsilon)\,dz \\ &\ge\; \int_{\mathcal{A}(s,\varepsilon)} \text{meas}\,\mathcal{B}(z,s,\varepsilon)\,dz \;\ge\; \rho_s\;\text{meas}\,\mathcal{A}(s,\varepsilon)\,. \end{aligned} \tag{3.3.15}$$

Since $\Phi'_\alpha(z,\alpha) = y - h'_\alpha(t,y)$ and $\Phi''_{\alpha\alpha}(z,\alpha) \;=\; -h''_{\alpha\alpha}(t,\alpha) \;=\; -D_*(t,\alpha)$, one has

$$\text{meas}\,\mathcal{B}(s,\varepsilon) \;\le\; 4\;\delta_s\;\text{meas}\,\mathrm{A}(s^2) \tag{3.3.16}$$

where $\mathrm{A}(\sigma) := \{(t,\alpha) \in \Omega_{\underline{t}} \times \mathcal{I}\;:\;|D_*(t,\alpha)|^{-1} \ge \sigma\}$. Therefore, (3.3.15) implies

$$\text{meas}\,\mathcal{A}(s,\varepsilon) \;\le\; \frac{4\,\delta_s}{\rho_s}\;\text{meas}\,\mathrm{A}(s^2) \;=\; \frac{4}{s^2}\;\text{meas}\,\mathrm{A}(s^2)\,. \tag{3.3.17}$$

6. If $D_*^{-1} \in L^{r,\infty}$, then $\operatorname{meas} \mathrm{A}(\sigma) \leq \Gamma\, \sigma^{-r}$. Thus $\operatorname{meas} \mathcal{A}(s,\varepsilon) \leq \Gamma\, s^{-(2+2r)}$ so

$$\sup_{\varepsilon \in]0,\varepsilon_0[} \| u^\varepsilon \|_{L^{2+2r,\infty}} < \infty . \tag{3.3.18}$$

Similarly, (3.3.17) implies that

$$\begin{aligned} \| u^\varepsilon \|_{L^{2+2r}} &\leq C \int_0^{+\infty} s^{1+2r} \operatorname{meas} \mathcal{A}(s,\varepsilon)\, ds \\ &\leq 4\, C \int_0^{+\infty} s^{2r-1} \operatorname{meas} \mathrm{A}(s^2)\, ds \leq C' \| D_*^{-r} \|_{L^1(\mathcal{O}_{\underline{t}} \times \mathcal{I})}, \end{aligned}$$

which completes the proof of Theorem 3.1.2.

Examples and Remarks 3.3.3. The assumptions of Theorem 3.1.2 are satisfied when

$$\Phi(t,y,\alpha) = y\,\alpha - \psi(t,\alpha), \qquad (t,y,\alpha) \in \mathbb{R}^{d-1} \times \mathbb{R} \times \mathbb{R}, \tag{3.3.19}$$

and ψ is a polynomial in α of degree $k \geq 2$, whose leading coefficient does not vanish for almost all t. In this case, $C_\Phi = \{y = \varphi'_\alpha(t,\alpha)\}$ and $\mathcal{S}_\Phi = C_\Phi \cap \{\psi''_{\alpha,\alpha}(t,\alpha) = 0\}$. No smoothness assumption is imposed on $\mathcal{S}_\Phi$.

The examples $\psi(t,\alpha) = \alpha^3$ and $\psi(t,\alpha) = t_1\, \alpha^2 + \alpha^4$, yield caustics which are folds and cusps. In these two examples the critical index q_c is equal to 4. This shows that the critical index q_c is not a function of the order of the caustic as introduced in [Du].

3.4. Proof of Theorem 3.1.3

Consider the oscillatory integrals (3.1.1), assuming that $\mathcal{S}_\Phi$ is a smooth manifold of C_Φ of codimension one, that the dimension ν of $\ker \Phi''_{\alpha\alpha}(\rho)$ is constant for $\rho \in \mathcal{S}_\Phi$, and that D_*, the restriction of $\det \Phi''_{\alpha\alpha}$ to C_Φ vanishes to order $\mu = \nu \geq 2$ on $\mathcal{S}_\Phi$.

After a linear change of variables, one has $\alpha = (\alpha', \alpha'') \in \mathbb{R}^\nu \times \mathbb{R}^{n-\nu}$, with $\Phi''_{\alpha',\alpha'}(\underline{\rho}) = 0$ and $\Phi''_{\alpha''\alpha''}(\underline{\rho})$ invertible. Using the stationary phase theorem in the α'' variables reduces the proof to the case $n = \nu$, which we assume from now on. This means that

$$\forall \rho \in \mathcal{S}_\Phi, \quad \Phi''_{\alpha\alpha}(\rho) = 0 . \tag{3.4.1}$$

Since the phase is non degenerate this implies that

$$\operatorname{rank} \Phi''_{z\alpha}(\underline{\rho}) = \nu. \tag{3.4.2}$$

In particular, $m \geq \nu$. After a linear change of coordinates, one can write $z = (t,x) \in \mathbb{R}^{m-\nu} \times \mathbb{R}^\nu$ with

$$\det \Phi''_{x\alpha} \neq 0 . \tag{3.4.3}$$

Considering t as a parameter, (3.4.3) implies that

$$C_\Phi(t) := \{\, (x,\alpha) \ : \Phi'_\alpha(t,x,\alpha) = 0 \,\}$$

is a smooth manifold of dimension ν. The associated Lagrangian manifold is

$$\Lambda_\Phi(t) := \{\, (x, d_x\Phi(t,x,\alpha)) : (x,\alpha) \in C_\Phi(t) \,\}.$$

Moreover, the mapping $(x, \xi) \to \xi$ is a local diffeomorphism from $\Lambda_\Phi(t)$ to $\mathbb{R}^\nu$. Therefore $\Lambda_\Phi(t)$ can be parametrized by ξ as the set $\{(d_\xi h(t,\xi), \xi)\}$, with a smooth function h defined near the point $(\underline{t}, \underline{\xi})$ corresponding to $\underline{\rho}$. Consider the phase function

$$\Psi(t, x, \xi) \;:=\; x \cdot \xi \;-\; h(t, \xi)\,. \tag{3.4.4}$$

The function h is uniquely determined if one requires that $\Phi(\underline{\rho}) = \Psi(\underline{t}, \underline{x}, \underline{\xi})$.

The theorem of equivalence of phase functions (see [Hö 1] [Du]), with t as parameters, implies that there exists a mapping $(z, \alpha) \mapsto \xi(z, \alpha)$ which is, for all fixed z, a local diffeomorphism $\alpha \mapsto \xi$ such that

$$\Phi(z, \alpha) \;=\; \Psi(z, \xi(z, \alpha))\,. \tag{3.4.5}$$

Changing variables implies that, for a supported in a sufficiently small neighborhood of $\underline{\rho}$,

$$u^\varepsilon(t, x) \;=\; \varepsilon^{-\nu/2} \int_{\mathbb{R}^\nu} e^{i\,\Psi(t,x,\xi)/\varepsilon}\, b(t, x, \xi, \varepsilon)\; d\xi\,, \tag{3.4.6}$$

where b is a smooth function on $U \times [0,1]$ and U a neighborhood of $(\underline{t}, \underline{x}, \underline{\xi})$.

The new phase Ψ defines the same Lagrangian manifold Λ as Φ. Thus Ψ satisfies the same assumptions as Φ. The manifold C_Ψ is given by the relation $x = h'_\xi(t,\xi)$ and is parametrized by (t, ξ). In this parametrization the manifold $\mathcal{S}_\Psi$ is locally given by an equation

$$f(t, \xi) = 0 \quad \text{with} \quad df(\underline{t}, \underline{\xi}) \neq 0\,. \tag{3.4.7}$$

Then (3.4.1) means that

$$h''_{\xi\xi}(t, \xi) = 0 \quad \text{when} \quad f(t, \xi) = 0\,. \tag{3.4.8}$$

In addition, the condition that D_* vanishes exactly to order μ on $\mathcal{S}_\Phi$ means that

$$\det h'' = f^\mu\, e \quad \text{with} \quad e(\underline{t}, \underline{x}, \underline{\xi}) \neq 0\,. \tag{3.4.9}$$

LEMMA 3.4.1. *There is a neighborhood ω of $(\underline{t}, \underline{x}, \underline{\xi})$ and a C^∞ function $g(t)$ near $\underline{t}$, such that $g(\underline{t}) = 0$, $dg(\underline{t}) \neq 0$ and*

$$h''_{\xi\xi}(t, \xi) \;=\; g(t)^r\, E(t, \xi)\,, \qquad \textit{with} \quad \det E(t, \xi) \;\neq\; 0\,. \tag{3.4.10}$$

Proof. (3.4.8) implies that there is a smooth symmetric matrix $E(t, x)$ with

$$h''_{\xi\xi}(t, \xi) \;=\; f(t, \xi)\, E(t, \xi)\,. \tag{3.4.11}$$

Since $\mu = \nu$, (3.4.9) implies that $\det E(t, \xi) \neq 0$. We show that (3.2.13) implies that $d_\xi f = 0$ when $f = 0$. This implies that $\mathcal{S}_\Psi := \{f = 0\}$ is also given by an equation $g(t) = 0$ and the proposition then follows.

Differentiating (3.4.11) with respect to ξ shows that

$$h'''_{\xi\xi\xi}(t, \xi)(\delta\xi_1, \delta\xi_2, \delta\xi_3) \;=\;$$
$$f'_\xi(t, \xi)\,(\delta\xi_3)\, E(t, \xi)\,(\delta\xi_1, \delta\xi_2) + f(t, \xi)\, E'_\xi(t, \xi)\,(\delta\xi_3, \delta\xi_1, \delta\xi_2)\,.$$

The left hand side is symmetric with respect to $(\delta\xi_1, \delta\xi_2, \delta\xi_3)$. Therefore, where $f(t,\xi) = 0$,

$$f'_\xi(t,\xi)\,(\delta\xi_3)\,E(t,\xi)\,(\delta\xi_1,\delta\xi_2) \;=\; f'_\xi(t,\xi)\,(\delta\xi_1)\,E(t,\xi)\,(\delta\xi_2,\delta\xi_3)\,. \tag{3.4.12}$$

Since $\nu \geq 2$, there is a $\delta\xi_1 \neq 0$ such that $f'_\xi(t,\xi)(\delta\xi_1) = 0$. Since $E(t,\xi)$ is symmetric and nondegenerate, there is $\delta\xi_2$ such that $E(t,\xi)\,(\delta\xi_1,\delta\xi_2), \neq 0$. Therefore (3.4.12) implies that $f'_\xi(t,\xi)\,(\delta\xi_3) = 0$ for all $\varphi\xi_3$. This shows that $d_\alpha f(t,\xi) = 0$ when $f(t,\xi) = 0$ and the lemma is proved.

Changing Ψ to $\Psi(t,x,\eta) + \psi(t,x)$ does not change the modulus $|u^\varepsilon|$ of the oscillatory integrals (3.4.6). Thus, we can also assume that $\underline{\eta} = 0$ and that $h(t,0) = 0$. Therefore, for η in a neighborhood of 0, one has

$$h(t,\eta) \;=\; h'_\eta(t,0)(\eta) \;+\; \int_0^1 (1-s)\; h''_{\eta\eta}(t,s\eta)\,(\eta\,,\,\eta)\;ds\,.$$

The change of variables $(t,x) \mapsto (t, x - h'_\eta(t,0))$ reduces the proof of Theorem 3.1.3 to the case where $h'_\eta(t,0) = 0$. In this case, (3.4.10) implies that

$$h(t,\eta) \;=\; g(t)\,\psi(t,\xi)\,, \qquad \text{with} \quad \det\psi''_{\xi\xi}(t,0) \neq 0\,. \tag{3.4.13}$$

In addition, $dg(\underline{t}) \neq 0$. All the reductions above being performed, Theorem 3.1.3 is a consequence of the following estimate.

PROPOSITION 3.4.2. *Consider the phase function* (3.4.4) *with h satisfying* (3.4.13). *There is a neighborhood ω of $\underline{\rho} = (\underline{t},\underline{x},\underline{\xi})$, such that for all $b \in C_0^\infty(\omega \times [0,1])$, there is a constant C such that the oscillatory integrals* (3.4.6) *satisfy*

$$|\,u^\varepsilon(t,x)\,| \;\leq\; C\;(|g(t)| \,+\, |x|)^{-\nu/2}\,. \tag{3.4.14}$$

Proof. Introduce the oscillatory integrals

$$v^\delta(t,y,x) \;:=\; \delta^{-\nu/2} \int e^{i\,(y\cdot\xi \,-\, \psi(t,\xi))/\delta}\; b(t,x,\xi)\;d\xi\,. \tag{3.4.15}$$

There are $R > 0$ and $c > 0$ such that for $|y| \geq R$, $|d_\xi(y\cdot\xi - \psi(t,\xi)| \geq c|y|$ on the support of b. Integrations by parts show that for all N there is C_N such that

$$\forall |y| \geq R\,,\; \forall \delta \leq |y|\,, \quad |\,v^\delta(t,y,x)\,| \;\leq\; C_N(|\delta|^{N-\nu/2}\,|y|^{-N})\,. \tag{3.4.16}$$

Since $\psi''_{\eta\eta}$ is nondegenerate at the origin, if the support of b is sufficiently small, then v^δ is uniformly bounded for bounded y and bounded δ, that is there is C such that

$$\forall |y| \leq R\,,\; \forall |\delta| \leq 1\,, \quad |\,v^\delta(t,y,x)\,| \;\leq\; C\,. \tag{3.4.17}$$

Moreover, v^δ is compactly supported in (t,x).

For $g(t) \neq 0$, one has

$$u^\varepsilon(t,x) \;=\; g(t)^{-\nu/2}\; v^\delta(t,\, x/g(t),\, x)\,, \qquad \text{with} \quad \delta = \varepsilon/g(t)\,. \tag{3.4.18}$$

1. For $|x| \le R\,|g(t)|$ and $\varepsilon \le |g(t)|$, (3.4.17) shows that $u^\varepsilon = O(|g(t)|^{-\nu/2}) = O\big((|g(t)| + |x|)^{-\nu/2}\big)$.

2. For $|x| \ge R\,|g(t)|$ and $\varepsilon \le |x|$, (3.4.16) implies that $u^\varepsilon = O(\varepsilon^{N-\nu/2}|x|^{-N})$. Taking $N \ge \nu/2$, this implies that $u^\varepsilon = O(|x|^{-\nu/2}) = O\big((|g(t)| + |x|)^{-\nu/2}\big)$. This estimate extends to the case $g(t) = 0$, since then the phase has no critical points.

3. In the other cases, one has $|g(t)| + |x| \le 2\varepsilon$, and $u^\varepsilon = O(\varepsilon^{-\nu/2}) = O\big((|g(t)| + |x|)^{-\nu/2}\big)$.

Summarizing, we have proved that when the support of b is contained in a sufficiently small neighborhood of $(\underline{t}, \underline{x})$, the estimate (3.4.14) holds and the proof is complete.

EXAMPLE 3.4.3. Spherical focusing leads to the phases $\Phi(t, x, \xi) = x\cdot\xi - t\,|\xi|^2$, for $(t, x, \xi) \in \mathbb{R} \times \mathbb{R}^\nu \times \mathbb{R}^\nu$ which are covered by these estimates.

3.5. Subcritical L^p estimates for $I^\varepsilon(\mathcal{A})$

This section is devoted to the proof of Theorem 2.3.5. Recall several notations and assumptions from §2. Consider oscillatory integrals of the form

$$(3.5.1)\qquad I^\varepsilon(\mathcal{A})\,(t,x) \;:=\; \sum_n \,(|n|/2\pi\varepsilon)^d \int e^{i\,n\,\Phi(t,x,y,\xi)/\varepsilon}\, a_n(t,y)\; dy\, d\xi\,,$$

with

$$(3.5.2)\qquad \Phi(t,x,y,\xi) := t\,\lambda(\xi) \,+\, (x-y)\cdot\xi \,+\, \psi(y)\,,$$

where ψ satisfies Assumptions 2.2.1 on $\omega \subset \mathbb{R}^d$, $\lambda(\xi)$ is a C^∞ function on $\mathbb{R}^d\backslash\{0\}$ homogeneous of degree one and

$$(3.5.3)\qquad \mathcal{A}(t,y,\theta) \;:=\; \sum_{n\neq 0} a_n(t,y)\; e^{in\theta}\;, \qquad a_n \in C_0^\infty([0,+\infty[\times\omega)$$

is a trigonometric polynomial. This is a particular case of integrals (3.1.1) with space variables $z = (t,x)$ and phase variables $\alpha = (y,\xi)$ except that the domain of integration is not compact in ξ.

The phase Φ is nondegenerate and the set C_Φ is parametrized by $\mathbb{R}\times\omega$ as follows

$$(3.5.4)\quad \iota\,:\,(t,y) \mapsto (t, y - t\,g(y), \lambda(d\psi(y)), d\psi(y))\,, \quad \text{with} \quad g(y) := \lambda'(d\psi(y))\,.$$

Then,

$$(3.5.5)\qquad \det \Phi''_{\alpha\alpha}(\iota(t,y)) \;=\; \det(I - t\,g'(y))\,,$$

so the singular set $\mathcal{S}_\Phi$ is equal to $\iota(\mathcal{S})$ where

$$(3.5.6)\qquad \mathcal{S} \;:=\; \big\{\,(t,y)\;:\;\det(I - t\,g'(y)) = 0\,\big\}\,.$$

The Caustic set is $\mathcal{C} := \jmath(\mathcal{S})$ where $\jmath(t,y) := (t,x)$ with $x = q(t,y) := y - tg(y)$. (3.5.6) shows that $(t,y)\in\mathcal{S}$ if and only if $1/t$ is an eigenvalue of $g'(y)$. Following Definition 2.3.2 denote by $\mu(t,y)$ the algebraic multiplicity of the eigenvalue and by $\nu(t,y)$ the dimension of the eigenspace. The multiplicity μ is constant on a neighborhood of (t,y) in $\mathcal{S}$ if and only if the eigenvalue has constant multiplicity on a neighborhood of y. The next lemma is then immediate.

LEMMA 3.5.1. *Consider* $(\underline{t}, \underline{y}) \in \mathcal{S}$. *Suppose that* μ *and* ν *are constant on a neighborhood of* $(\underline{t}, \underline{y})$ *in* $\mathcal{S}$. *Then, on a neighborhood of* $(\underline{t}, \underline{y})$, $\mathcal{S}$ *is a smooth manifold of codimension one and the determinant* $\det(I - t\, g'(y))$ *vanishes exactly to order* μ *on* $\mathcal{S}$.

REMARKS AND EXAMPLES 3.5.2. a) One has $1 \le \nu \le \mu \le d-1$, see (2.6.1). In particular, in space dimension $d = 2$, the only possibility is $\nu = \mu \doteq 1$.

b) The condition $\mu = \nu$ means that the corresponding eigenvalue is semi-simple. If either λ_k'' or ψ'' is positive (or negative) semidefinite, then the nonzero eigenvalues of $g_k'(y) = \lambda''(d\psi(y))\, \psi''(y)$ are semi-simple.

This applies in particular to the wave equation, where $\lambda_k(\xi) = \pm|\xi|$ is convex (concave). In this case, the eigenvalues of $g_k'(\underline{y})$ are 0 and $\pm$ the principal curvatures of the manifold $\psi(y) = \psi(\underline{y})$.

The same remark applies to the Maxwell equations, the linearized Euler equations, and linear elasticity for each of which each eigenvalue defines a globally convex or concave function.

c) Generic caustics occur when $\nu = \mu = 1$, and $Vc \ne 0$ where $V(y)$ is a smooth nonvanishing eigenvectorfield $V(y) \in \ker(c(y)I - g'(y))$ where $c(y)$ is the eigenvalue of $g'(y)$ under consideration (see [Hö 2][Du][JMR 3]). In this case, $\mathcal{C}$ is smooth and $\pi_{|\Lambda}$ is a fold.

d) When $\nu = \mu = 1$ and $Vc(\underline{y}) = 0$, $V^2 c(\underline{y}) \ne 0$, the caustic $\mathcal{C}$ has a cusp. This applies to the wave equation in $\mathbb{R}^{1+2}$ and the phases (2.6.4), where only folds and cusps are present.

e) Remarks (c) and (d) applies to the wave equation in $\mathbb{R}^{1+2}$ and the phases (2.6.4). Points where the curvature is maximum or minimum give cusps. Points where the curvature is not stationary give folds.

f) For the wave equation and radial phases $\psi(y) = \chi(|y|)$, there is one positive eigenvalue, $c(y) = |y|$, semisimple and of multiplicity $\mu = \nu = d - 1$.

The Assumptions 2.2.2 and 2.3.3 read

ASSUMPTION 3.5.3. *i) There is an integer* ℓ_*, *such that for and all* $(t, x) \notin \mathcal{C}$, *the number of points* $y \in \omega$ *such that* $x = q(t, y)$, *is less than or equal to* ℓ_*.

ii) There is an open subset $\omega_0 \subset \omega$ *such that* $\omega \backslash \omega_0$ *has Lebesgue measure equal to 0 and for all* k *and all* $(t, y) \in \mathcal{S}$ *with* $y \in \omega_0$, *one has*

a) μ *and* ν *are constant on a neighborhood of* (t, y) *in* $\mathcal{S}$,

b) either $\nu = 1$ *or* $\nu = \mu$.

REMARK 3.5.4. a) When λ and ψ are real analytic, the phase Φ is real analytic and the Assumption *i)* is satisfied.

b) The eigenvalues of $g'(y)$ are locally of constant algebraic and geometric multiplicity on a dense open subset of ω. Part a) of Assumption *ii)* says that this set has full measure in ω. In particular, this is true when λ and ψ are real analytic.

c) Combined with Remarks 3.5.2 a) and b), this shows that Assumption 3.5.3 is satisfied in the following three cases.

1) Real analytic λ and real analytic convex phases ψ.
2) Real analytic phases for the wave equations.
3) Real analytic λ and real analytic phases in space dimension $d = 2$.

d) Assumption 3.5.3 is also satisfied in the example 6 of §2.6, where $\lambda_k(\xi) = |\xi|$ and

$$\psi(y) := y_d + (y_1^2 + \cdots + y_{d-1}^2)/2\,, \tag{3.5.7}$$

with $\omega_0 := \{y' := (y_2, \ldots, y_d) \neq 0\}$.

For $q \in [2, \infty]$, the set of supercritical points $\mathcal{S}^+(q)$ is the set of $(t, y) \in \mathcal{S}$ such that either $y \in \omega \backslash \omega_0$ or $\mu(t, y) \geq 2/(q-2)$. Theorem 2.3.5 is an immediate corollary of the following result.

THEOREM 3.5.5. *Suppose that Assumption 3.5.3 is satisfied. Let $q \geq 2$. Assume that the coefficients a_n of $\mathcal{A}$ are compactly supported in $[0, +\infty[\times\omega$ and vanish on a neighborhood of the points $(t, y) \in \mathcal{S}^+(q)$. Then,*

$$\sup_{\varepsilon \in]0,1]} \|\, I^\varepsilon(\mathcal{A})\, \|_{L^q(\mathbb{R}^{1+d})} < +\infty\,. \tag{3.5.8}$$

Proof. a) It is sufficient to consider the case where $\mathcal{A}$ is a monomial. Changing ε to $\varepsilon/|n|$ and taking complex conjugate it is sufficient to consider integrals of the form

$$u^\varepsilon(t, x) := \varepsilon^{-d} \int a(t, y)\, e^{i\,\Phi(t,x,y,\xi)/\varepsilon}\, dy\, d\xi\,, \tag{3.5.9}$$

b) Choose $\chi \in C^\infty(\mathbb{R}^d)$ such that $1 - \chi$ has compact support disjoint from the origin and $\chi = 0$ on a neighborhood of the support of $d\psi$. Let

$$u^\varepsilon_\chi(t, x) := \varepsilon^{-d} \int a(t, y)\, \chi(\xi)\, e^{i\,\Phi(t,x,y,\xi)/\varepsilon}\, dy\, d\xi\,. \tag{3.5.10}$$

Since $|\partial_y \Phi| \geq c(1 + |\xi|)$ on the support of $a(t, y)\,\chi(\xi)$, integration by parts in y yields $u^\varepsilon_\chi = O(\varepsilon^\infty)$ uniformly in (t, x).

To estimate the L^2 norm, write

$$u^\varepsilon_\chi(t\,.) = e^{i\,t\,\lambda(D)}\, \chi(\varepsilon\, D)\, \big(e^{i\,\psi/\varepsilon}\, a(t,\,.)\big)\,.$$

The $L^2(\mathbb{R}^d)$ norm is the same as that without the factor $e^{i\,t\,\lambda(D)}$. The expression (3.5.10) without λ has phase $\widetilde{\Phi} = (x - y)\cdot\xi + \psi(y)$ which satisfies $|\partial_y \widetilde{\Phi}| \geq c(1 + |\xi|)$ on the support of χ and $|\partial_\xi \widetilde{\Phi}| \geq c|x|$ on the support of a if $|x|$ is large. So, integration by parts in (y, ξ) yields

$$\big|\, (e^{-i\,t\,\lambda(D)}\, u^\varepsilon_\chi)\,(t, x)\, \big| \leq O(\,\varepsilon^N/(1 + |x|)^N)\,,$$

for all N. Therefore $u^\varepsilon_\chi(t,\,.) = O(\varepsilon^\infty)$ in L^2. This proves that for all $p \geq 2$,

$$\|\, u^\varepsilon_\chi(t,\,\cdot)\|_{L^p(\mathbb{R}^d)} = O(\varepsilon^\infty)\,. \tag{3.5.11}$$

c) Consider

$$u^\varepsilon_{1-\chi}(t,x) \;:=\; \varepsilon^{-d}\int a(t,y)\,(1-\chi(\xi))\,e^{i\,\Phi(t,x,y,\xi)/\varepsilon}\,dy\,d\xi\,. \tag{3.5.12}$$

Since a is compactly supported in y, $|d_\xi\Phi| \geq c(|x|+1)$ when $|x|$ is larger than R. Therefore, $u^\varepsilon_{1-\chi}$ is $O((\varepsilon/|x|)^\infty)$ for $|x| \geq R$. Therefore

$$\|\,u^\varepsilon_{1-\chi}(t,\,\cdot\,)\|_{L^p(|x|\geq R)} \;=\; O(\varepsilon^\infty)\,. \tag{3.5.13}$$

d) Thus, it is sufficient to bound the L^q norms of $u^\varepsilon_{1-\chi}$ on compact sets in $[0,+\infty[\times\mathbb{R}^d$. The nonstationary points contribute $O(\varepsilon^\infty)$, and the nondegenerate stationary points contributes $O(1)$. Therefore, using a partition of unity, it is sufficient to prove that the L^q norm of $u^\varepsilon_{1-\chi}$ is bounded independently of ε when a is supported in a small neighborhood of $(\underline{t},\underline{y}) \in \mathcal{S}$ and $(1-\chi)$ is supported in a small neighborhood of $\underline{\xi} := d\psi(\underline{y})$. Since the original a is supported away from $\mathcal{S}^+(q)$ it is sufficient to consider points $(\underline{t},\underline{y}) \in \mathcal{S}$ such that $\underline{y} \in \omega_0$ and $\mu(\underline{t},\underline{y}) < 2/(q-2)$. In this case, Assumptions 3.5.3 implies that either Theorem 3.1.2 or Theorem 3.1.3 applies (see Remark 3.1.4) and Theorem 3.5.5 follows.

REMARK 3.5.6. If the hypothesis of Theorem 3.5.5 is satisfied for $q \geq 4$ then the conclusion 3.5.8 can be strengthened. In fact, 3.5.8 then holds for all $q \geq 2$. To see this note that $2/(q-2) \leq 1$ for $q \geq 4$, which implies that all points in $\mathcal{S}$ are supercritical. Thus the hypothesis implies that $\mathcal{A}$ vanishes for (t,y) in a neighborhood of $\mathcal{S}$. Therefore the critical points of the phase Φ_k located in the support of the coefficients are nondegenerate, so the family $I^\varepsilon(\mathcal{A})$ is bounded in L^∞. With L^2 estimates provided by the Plancherel identity, this implies that (3.5.8) holds for all $q \geq 2$.

On the other hand, when $q < 2 + 2/(d-1)$, (3.5.8) holds for all trigonometric polynomials with coefficients in $C_0^\infty([0,+\infty[\times\omega_0)$, since the multiplicity μ is at most $d-1 < 2/(q-2)$. Thus the index $2+2/(d-1)$ which occurs for spherical focusing is the worst case.

4. Oscillations and profiles

In the first part of this section, we introduce families $I^\varepsilon(\mathcal{A})$ and $J^\varepsilon(\mathcal{A})$. They are defined in (2.3.2) and (2.4.6) when $\mathcal{A}$ is a smooth trigonometric polynomial. The definition of J^ε is extended to $\mathcal{L}_k^{s,q}$ for $s > 1/q$ and also, in an asymptotic sense, to $\mathcal{L}_k^q$. In the second part of the section, we introduce profiles associated to families u^ε. This is a modification of the multiscale analysis introduced in [N], [A], [E], [ES], [JMR 2] with phases replaced by Lagrangians.

In this section we suppose that Assumptions 2.1.1, 2.2.1, 2.2.2, 2.2.5 and 2.3.3 are satisfied.

4.1. Estimates for $J^\varepsilon(\mathcal{A})$ and $I^\varepsilon(\mathcal{A})$

DEFINITION 4.1.1. *$\mathcal{P}_k \subset \mathcal{P}$ denotes the space of trigonometric polynomials whose coefficients vanish in a neighborhood of $\mathcal{S}_k$. $\mathcal{P}_k(q)$ denotes the space of trigonometric polynomials which vanish on a neighborhood of the set of supercritical point $\mathcal{S}_k^+(q)$ introduced in Definition 2.3.4.*

For $\mathcal{A} \in \mathcal{P}_k$ the family $J_k^\varepsilon(\mathcal{A})$ is defined by (2.4.2).

PROPOSITION 4.1.2. *For $\mathcal{A} \in \mathcal{P}_k$, and $\varepsilon > 0$, $J_k^\varepsilon(\mathcal{A}) \in C_0^\infty([0,+\infty[\times\mathbb{R}^d)$. For each $q \in]1,+\infty[$, there is a constant C such that for all $\mathcal{A} \in \mathcal{P}_k$,*

$$\limsup_{\varepsilon \to 0} \; ||\, J_k^\varepsilon(\mathcal{A})\,||_{L^q([0,\infty[\times\mathbb{R}^d)} \;\le\; C \;\; ||\,\mathcal{A}\,||_{\mathcal{L}_k^q} \, . \tag{4.1.1}$$

For $1/q < s < 1$, there is a constant C such that for all $\mathcal{A} \in \mathcal{P}_k$,

$$\sup_{0<\varepsilon\le 1} \;\; ||\, J_k^\varepsilon(\mathcal{A})\,||_{L^q([0,\infty[\times\mathbb{R}^d)} \;\le\; C \;\; ||\,\mathcal{A}\,||_{\mathcal{L}_k^{s,q}} \, . \tag{4.1.2}$$

Proof. **a)** The support of $\mathcal{A} \in \mathcal{P}_k$ is contained in $K \times \mathbb{T}$ where K is a compact subset of G which does not intersect $\mathcal{S}_k$. Therefore, there is a covering of K by relatively open sets $G_\alpha \subset G$, such that $\jmath_k$ is a diffeomorphism from G_α onto $\Omega_\alpha := \jmath_k(G_\alpha)$. Denote by σ_α the inverse mapping from Ω_α onto G_α. Introduce a finite partition of unity, $\chi_\alpha \in C_0^\infty(G_\alpha)$, such that $\chi_\alpha \ge 0$ and $\sum \chi_\alpha = 1$ on K. Then $\mathcal{A} = \sum \chi_\alpha \mathcal{A}$, $J_k^\varepsilon(\mathcal{A}) = \sum J_k^\varepsilon(\chi_\alpha \mathcal{A}_\alpha)$ and

$$J_k^\varepsilon(\chi_\alpha \mathcal{A})\,(t,x) \;=\; \frac{\chi_\alpha(\sigma_\alpha(t,x))}{\Delta_k(\sigma_\alpha(t,x))}\;(\mathcal{H}^{m_{k,\alpha}}\mathcal{A})\,(\sigma_\alpha(t,x), \varphi_\alpha(t,x)/\varepsilon)\,, \tag{4.1.3}$$

where $\varphi_\alpha(t,x) := \psi(\sigma_\alpha(t,x))$ and $m_{k,\alpha}$ is the value of $m_k(t,y)$ on G_α. This proves that $J_k^\varepsilon(\mathcal{A})$ is C^∞ and has compact support contained in $\jmath_k(K)$.

b) The finiteness Assumption 2.2.2 implies that

$$| J_k^\varepsilon(\mathcal{A})\,(t,x)\,|^q \;\le\; \ell_*^{q-1} \sum_{\{y \,|\, q_k(t,y)=x\}} \frac{1}{|\Delta_k(t,y)|^q}\; |\, (\mathcal{H}^{m_k(t,y)}\mathcal{A})\,(t,y,\psi(y)/\varepsilon)\,|^q .$$

Introduce $\mathcal{B} := \mathcal{H}^{m_k}\mathcal{A}$. $\mathcal{B}$ is smooth in t, y since m_k is locally constant on the support of $\mathcal{A}$ and therefore $\mathcal{B} \in \mathcal{P}_k$. The identity $|\mathcal{B}|^q = \sum \chi_\alpha |\mathcal{B}|^q$ where χ_α is the partition of unity introduced in a), implies that

$$| J_k^\varepsilon(\mathcal{A})\,(t,x)\,|^q \;\le\; \ell_*^{q-1} \sum_\alpha \frac{\chi_\alpha(\sigma_\alpha(t,x))}{|\Delta_k(\sigma_\alpha(t,x))|^q}\; |\, \mathcal{B}\,(\sigma_\alpha(t,x), \varphi_\alpha(t,x)/\varepsilon)\,|^q .$$

Since Δ_k^2 is the Jacobian of q_k, one has

$$\text{(4.1.4)} \qquad \|\, J_k^\varepsilon(\mathcal{A})\,\|_{L^q}^q \;\le\; \ell_*^{q-1} \sum_\alpha \int_{G_\alpha} \frac{\chi_\alpha(t,y)}{|\Delta_k(t,y)|^{q-2}}\; |\, \mathcal{B}\,(t,y,\psi(y)/\varepsilon)\,|^q \; dtdy\,.$$

c) $|\mathcal{B}(t,y,\theta)|^q$ is a compactly supported and continuous function. Fourier expansion in θ yields

$$\lim_{\varepsilon\to 0} \int_{G_\alpha} \frac{\chi_\alpha(t,y)}{|\Delta_k(t,y)|^{q-2}}\; |\, \mathcal{B}\,(t,y,\psi(y)/\varepsilon)\,|^q \; dtdy = \int_{G_\alpha\times\mathbb{T}} \frac{\chi_\alpha(t,y)}{|\Delta_k(t,y)|^{q-2}}\; |\, \mathcal{B}\,(t,y,\theta)\,|^q \; dtdyd\theta$$

Therefore

$$\text{(4.1.5)} \qquad \limsup_{\varepsilon\to 0} \|\, J_k^\varepsilon(\mathcal{A})\,\|_{L^q}^q \;\le\; \ell_*^{q-1} \int |\Delta_k(t,y)|^{2-q}\; |\, \mathcal{B}\,(t,y,\theta)\,|^q \; dtdyd\theta\,.$$

For $q \in]1,+\infty[$, the Hilbert transform is bounded in $L^q(\mathbb{T})$. Thus

$$\int |\, \mathcal{B}\,(t,y,\theta)\,|^q \; d\theta \;\le\; C_q \int |\, \mathcal{A}\,(t,y,\theta)\,|^q \; d\theta\,.$$

With (4.1.5) this implies (4.1.1).

d) For $s > 1/q$, use the Sobolev inequality

$$\sup_{\theta\in\mathbb{T}} \; |\, \mathcal{B}(t,y,\theta)\,| \;\le\; C\, \|\, \mathcal{B}(t,y,\,.\,)\,\|_{W^{s,q}(\mathbb{T})}\,.$$

By definition of $\mathcal{L}_k^{s,q}$, there is a constant C such that

$$\text{(4.1.6)} \qquad \sup_{\theta\in\mathbb{T}} \; |\, \mathcal{B}(t,y,\theta)\,|^q \;\le\; Cb(t,y)$$

with

$$\text{(4.1.7)} \qquad \int_G \Delta_k^{2-q}\; b(t,y)\, dt\, dy \;\le\; \|\, \mathcal{B}\,\|_{\mathcal{L}_k^{s,q}}^q \;\le\; C\,\|\, \mathcal{A}\,\|_{\mathcal{L}_k^{s,q}}^q\,,$$

where the second inequality uses the fact that the Hilbert transform is bounded in $W^{s,q}(\mathbb{T})$. Plugging (4.1.6) in (4.1.4), and using (4.1.7) yields (4.1.2) and Proposition 4.1.2 is proved.

PROPOSITION 4.1.3. *i) For all trigonometric polynomials $\mathcal{A}$,*

$$\limsup_{\varepsilon\to 0} \| I_k^\varepsilon(\mathcal{A})(t,\cdot)\|_{L^2(\mathbb{R}^d)} = \| \mathcal{A}(t,\cdot)\|_{L^2(\omega\times\mathbb{T})} . \tag{4.1.8}$$

The limit is uniform on bounded intervals of time. In addition, for all $T>0$

$$\lim_{\varepsilon\to 0} \| I_k^\varepsilon(\mathcal{A}) - J_k^\varepsilon(\mathcal{A})\|_{L^2([0,T]\times\mathbb{R}^d)} = 0 . \tag{4.1.9}$$

ii) If $q\in[2,+\infty[$, there is a constant C such that for all trigonometric polynomials $\mathcal{A}\in\mathcal{P}_k(q)$,

$$\limsup_{\varepsilon\to 0} \| I_k^\varepsilon(\mathcal{A})\|_{L^q([0,T]\times\mathbb{R}^d)} \leq C \,\|\mathcal{A}\|_{\mathcal{L}_k^q} , \tag{4.1.10}$$

and

$$\lim_{\varepsilon\to 0} \| I_k^\varepsilon(\mathcal{A}) - J_k^\varepsilon(\mathcal{A})\|_{L^q([0,\infty[\times\mathbb{R}^d)} = 0 . \tag{4.1.11}$$

Proof. **a)** Part *i)* is a consequence of the standard L^2 estimates for oscillatory integrals (2.3.2). For a detailed proof, see [JMR 3].

b) For *ii)*, estimates (3.5.11) (3.5.13) imply that when $\mathcal{A}\in\mathcal{P}_k(q)$ the L^q norm of $I_k^\varepsilon(\mathcal{A})$ outside a compact subset K in $[0,+\infty[\times\mathbb{R}^d$ is $O(\varepsilon^\infty)$. Therefore, it is sufficient to prove that

$$\limsup_{\varepsilon\to 0} \| I_k^\varepsilon(\mathcal{A})\|_{L^q(K)} \leq C\,\|\mathcal{A}\|_{\mathcal{L}_k^q} . \tag{4.1.12}$$

c) Since $\mathcal{A}$ vanishes near points in $\mathcal{S}_k^+(q)$, points (t,y) in $\mathcal{S}_k\cap\operatorname{supp}\mathcal{A}$ satisfy $\mu_k(t,y)<2/(q-2)$. By compactness, one can find $r>q$ such that $\mu_k(t,y)<2/(r-2)$ and hence $\mathcal{A}\in\mathcal{P}_k(r)$. Therefore, Theorem 2.3.5 implies that $I_k^\varepsilon(\mathcal{A})$ is bounded in $L^r([0,+\infty[\times\mathbb{R}^d)$. Thus, there exists M such that for all bounded domains D,

$$\| I_k^\varepsilon(\mathcal{A})\|_{L^q(D)} \leq M\,(\text{meas}D)^{1/q-1/r} . \tag{4.1.13}$$

d) Since the caustic set $\mathcal{C}_k$ has Legesgue measure equal to zero, (4.1.13) implies that in order to prove (4.1.12), it is sufficient to show that for all bounded open sets Ω such that $\overline{\Omega}\cap\mathcal{C}_k=\emptyset$

$$\limsup_{\varepsilon\to 0} \| I_k^\varepsilon(\mathcal{A})\|_{L^q(\Omega)} \leq C\,\|\mathcal{A}\|_{\mathcal{L}_k^q} . \tag{4.1.14}$$

On such domains, the stationary phase theorem applies, and $I_k^\varepsilon(\mathcal{A})-J_k^\varepsilon(\mathcal{A})=O(\varepsilon)$ in L^∞. Therefore (4.1.10) follows from Proposition 4.1.2.

e) Similarly, Proposition 4.1.2 implies that $J_k^\varepsilon(\mathcal{A})$ satisfies estimates analogous to (4.1.13). Since $J_k^\varepsilon(\mathcal{A})$ is compactly supported, to prove (4.1.11), it is sufficient to show that

$$\lim_{\varepsilon\to 0} \| I_k^\varepsilon(\mathcal{A}) - J_k^\varepsilon(\mathcal{A})\|_{L^q(\Omega)} = 0 , \tag{4.1.15}$$

for all bounded open sets Ω such that $\overline{\Omega}\cap\mathcal{C}_k=\emptyset$. For such Ω, $I_k^\varepsilon(\mathcal{A})-J_k^\varepsilon(\mathcal{A})=O(\varepsilon)$ in L^∞, implying (4.1.15). The proof of Proposition 4.2.3 is complete.

PROPOSITION 4.1.4. *If $\mathcal{A}$ and $\mathcal{B}$ belong to $\mathcal{P}_k$ then, as $\varepsilon \to 0$,*

$$\int J_k^\varepsilon(\mathcal{A})\,\overline{J_k^\varepsilon(\mathcal{B})}\;dt\,dx \quad \to \quad \int \mathcal{A}\,\overline{\mathcal{B}}\;dt\,dy\,d\theta\,. \tag{4.1.16}$$

Proof. It is sufficient to prove (4.1.16) when $\mathcal{A} = a(t,y)\,e^{i\,n_\alpha\,\theta}$ and $\mathcal{B} = b(t,y)\,e^{i\,n_\beta\,\theta}$ where a and b have small supports. One can assume that they are supported in G_α and G_β respectively, where $\jmath_k$ is a diffeomorphism from both G_α and G_β onto a regular open set Ω. In this case, with notations as in (4.1.3),

$$J_k^\varepsilon(\mathcal{A})\,(t,x) \;=\; \frac{i^{m_{k,\alpha}}}{\Delta_k(\sigma_\alpha(t,x))}\;a\left(\sigma_\alpha(t,x)\right)\,e^{i\,n_\alpha\,\varphi_\alpha(t,x)/\varepsilon}\,, \tag{4.1.17}$$

and a similar formula for $J_k^\varepsilon(\mathcal{B})$. There are two cases.

1) $G_\alpha = G_\beta$, $\sigma_\alpha = \sigma_\beta$ and $\varphi_\alpha = \varphi_\beta$. If $n_\alpha = n_\beta$ the two sides of (4.1.16) are independent of ε and are equal since $\Delta_k^2(t,y)$ is the Jacobian of $\jmath_k$. If $n_\alpha \neq n_\beta$, the right hand side vanishes and the left hand side tends to 0, since $d\varphi_\alpha$ does not vanish.

2) $G_\alpha \cap G_\beta = \emptyset$. Then the integrand in the right hand side vanishes. The left hand side tends to 0, since $n_\alpha d\varphi_\alpha - n_\beta d\varphi_\beta \neq 0$ almost everywhere, thanks to the nonresonance Assumption 2.2.5.

4.2. Extension to nonsmooth profiles

PROPOSITION 4.2.1. *i) For all $q \in [1,\infty[$ and $s \in]0,1[$, $\mathcal{P}_k$ is dense in $\mathcal{L}_k^q$ and in $\mathcal{L}_k^{s,q}$.*

ii) For $1/q < s < 1$ and each $\varepsilon \in]0,1]$, J_k^ε extends to a bounded linear operator from $\mathcal{L}_k^{s,q}$ to $L^q([0,\infty[\times\mathbb{R}^d)$ and inequality (4.1.2) extends to $\mathcal{L}_k^{s,q}$.

Proof. Recall that $\mathcal{S}_k$ is the set of (t,y) such that $1/t$ is a positive eigenvalue of $g_k'(y)$. Thus its Lebesgue measure is zero, and the space of compactly supported functions which vanish on a neighborhood of $\mathcal{S}_k$ is dense in $\mathcal{L}_k^q$ and in $\mathcal{L}_k^{s,q}$ for all $s,q \in]0,1[\times[1,\infty[$. Using mollifiers, this implies that the space of compactly supported C^∞ functions on $[0,\infty[\times\omega\times\mathbb{T}$, which vanish on a neighborhood of $\mathcal{S}_k$, are dense. Truncating the Fourier series proves *i)*. Part *ii)* follows.

For general profiles, the definition of J^ε must be understood in an asymptotic sense, i.e. they are bounded families in a Banach space defined up to families which converge strongly to zero in the same space. A convenient setting is the following.

For a Banach space E, BE denotes the Banach space of bounded families $\{u^\varepsilon\}_{\varepsilon\in]0,1]}$ normed by $\sup \|u^\varepsilon\|_E$. NE is the closed subspace of families with $u^\varepsilon \to 0$ as $\varepsilon \to 0$. AE is the quotient space BE/NE. The norm in AE is equal to

$$\|\,[u^\varepsilon]\,\|_{AE} \;:=\; \limsup_{\varepsilon}\;\|\,u^\varepsilon\,\|_E\,. \tag{4.2.1}$$

The brackets to indicate equivalence classes in the following discussion is systematically suppressed. We write $u^\varepsilon \sim 0$ in E, whenever $u^\varepsilon \in NE$ and similarly $u^\varepsilon \sim v^\varepsilon$ in E when $u^\varepsilon - v^\varepsilon \in NE$.

PROPOSITION 4.2.2. *For all $q\in]1,\infty[$, the family of mappings J_k^ε defines a bounded mapping $\mathcal{J}_k$, from $\mathcal{L}_k^q$ to $AL^q([0,\infty[\times\mathbb{R}^d)$. It satisfies*

$$\frac{1}{C}\,||\,\mathcal{A}\,||_{\mathcal{L}_k^q}\;\le\;||\,\mathcal{J}_k(\mathcal{A})\,||_{L^q([0,\infty[\times\mathbb{R}^d)}\;\le\;C\;||\,\mathcal{A}\,||_{\mathcal{L}_k^q}\,. \tag{4.2.2}$$

Suppressing as always the quotients, the notation $\mathcal{J}_k(\mathcal{A})$ is replaced by $\tilde{J}_k^\varepsilon(\mathcal{A})$. Then

$$u^\varepsilon\;\sim\;\tilde{J}_k^\varepsilon(\mathcal{A})\quad\text{in } L^q\,, \tag{4.2.3}$$

means that the class of $[u^\varepsilon]$ of u^ε in $AL^q([0,\infty[\times(\mathbb{R}^d))$ is $\mathcal{J}_k(\mathcal{A})$. This is equivalent to saying that u^ε is a bounded family in $L^q([0,\infty[\times\mathbb{R}^d)$ such that for all $\delta>0$ there is an $\mathcal{A}\in\mathcal{P}_k$ and $\varepsilon_\delta>0$ such that

$$\begin{cases}||\,\mathcal{A}\,-\,\mathcal{A}_\delta\,||_{\mathcal{L}_k^q}\;\le\;\delta\,, & \text{and}\\[2mm] \forall\varepsilon\in]0,\varepsilon_\delta]\,,\quad ||\,u^\varepsilon\,-\,J_k^\varepsilon(\mathcal{A}_\delta)\,||_{L^q([0,\infty[\times\mathbb{R}^d)}\;\le\;\delta\,. & \end{cases} \tag{4.2.4}$$

Thus, Proposition 4.2.2 is a restatement of Propositions 2.4.2 and 2.4.3.

Proof. The extension and the second inequality follow from the density Lemma 4.2.1 and Proposition 4.1.2.

Propositions 4.1.2 and 4.1.4 imply that for trigonometric polynomials $\mathcal{A}$ and $\mathcal{B}$ in $\mathcal{P}_k$, one has

$$\Big|\int\mathcal{A}\,\overline{\mathcal{B}}\;dt\,dy\,d\theta\;\Big|\;\le C\;\liminf||\,J_k^\varepsilon(\mathcal{A})||_{L^q}\;||\,\mathcal{B}\,||_{\mathcal{L}_k^{q'}} \tag{4.2.5}$$

with $1/q+1/q'=1$. The density of $\mathcal{P}_k$ in $\mathcal{L}_k^{q'}$ and the converse Hölder inequality imply that for $\mathcal{A}\in\mathcal{P}_k$,

$$\frac{1}{C}\,||\,\mathcal{A}\,||_{\mathcal{L}_k^q}\;\le\;\liminf_{\varepsilon\to 0}\;||\,J_k^\varepsilon(\mathcal{A})\,||_{L^q([0,\infty[\times\mathbb{R}^d)}\,.$$

By density, this extends to $\mathcal{A}\in\mathcal{L}_k^q$ and the proposition is proved. This estimate also finishes the proof of Proposition 2.4.2.

There is a similar treatment for the Cauchy data, and the substitution

$$\mathcal{U}_0(y,\psi(y)/\varepsilon)$$

which is defined for $\mathcal{U}_0\in C^0(\omega\times\mathbb{T})$ with compact support and extends asymptotically from $L^2((\mathbb{R}^d\times\mathbb{T})$ to $AL^2(\mathbb{R}^d)$. The notation analogous to (4.2.3) is

$$u_0^\varepsilon\;\sim\;\tilde{J}_0^\varepsilon(\mathcal{U}_0)\quad\text{in } L^2(\mathbb{R}^d)\,.$$

REMARK 4.2.3. The operators J_k^ε are defined for fixed ε when $\mathcal{A} \in \mathcal{P}_k$ or more generally for $\mathcal{A} \in \mathcal{L}_k^{s,q}$, $s > 1/q$. In these cases, $J_k^\varepsilon(\mathcal{A})$ is a representative of $\mathcal{J}_k(\mathcal{A})$, that is,

$$J_k^\varepsilon(\mathcal{A}) \sim \tilde{J}_k^\varepsilon(\mathcal{A}).$$

It would be tempting to use (4.1.10) to extend I^ε asymptotically. By (4.1.11) this would yield $\tilde{I}^\varepsilon \sim \tilde{J}^\varepsilon$ and therefore nothing new. Note that the problems of extending I^ε and J^ε are different. For J^ε the difficulty is the substitution $\theta = \psi(y)/\varepsilon$ in (2.4.2). On the other hand, when $\mathcal{A}$ is smooth in θ, for example when $\mathcal{A}(t,y,\theta) := a(t,y)\, e^{i\,\theta}$, $I^\varepsilon(\mathcal{A})$ is defined for each ε, at least in the sense of distributions, but need not be bounded in L^q when $a \in L^q$. This is not due to the caustic but to the fact that Fourier Integral Operators of degree 0 are not bounded in L^q for $q \neq 2$. In particular, the limsup in (4.1.10) cannot be replaced by sup.

In the sequel, the operators I_k^ε act only on smooth profiles $\mathcal{A} \in \mathcal{P}_k(q)$. The $I_k^\varepsilon(\mathcal{A})$ serve as test functions.

4.3. Weak profiles

The analysis of weak convergence has been refined to a multiscale analysis, associating profiles to sequences of functions (see [N], [A], [E], [ES], and [JMR 2]). Given a phase φ, the idea is to test u^ε against functions of the form $\mathbf{B}(t,x,\varphi(t,x)/\varepsilon)$. Near a caustic there are several phases involved and the objects are better described by tests against functions of the form $J^\varepsilon(\mathcal{A})$.

DEFINITION 4.3.1. *If $1 < q < \infty$ and $\{u^\varepsilon\}_{0<\varepsilon<1}$ is a bounded family in $L^q(\mathbb{R}^{1+d})$, then $\mathcal{U} \in \mathcal{L}_k^q(G \times \mathbb{T})$ is called a Λ_k-profile of the family if there is a subsequence $u^{\varepsilon'}$ so that for all $\mathcal{A} \in \mathcal{L}_k^{q'}(G \times \mathbb{T})$ and all $v^\varepsilon \sim \tilde{J}_k^\varepsilon(\mathcal{A})$ in $L^{q'}$.*

$$\lim_{\varepsilon' \to 0} \int \overline{v^{\varepsilon'}}\; u^{\varepsilon'}\; dt\, dx \;=\; \int \overline{\mathcal{A}}\, \mathcal{U}_k\; dt\, dy\, d\theta\,. \tag{4.3.1}$$

We say that a sequence u^ε is k-pure when it converges weakly and the convergence (4.3.1) holds for the full family u^ε. It is pure, when it is k-pure for all k.

The estimate (4.1.1) and Lemma 4.2.1 show that it suffices to verify (4.3.1) for $\mathcal{A} \in \mathcal{P}_k$. The next proposition implies that any bounded family in L^q has pure subsequences.

PROPOSITION 4.3.2. *If $1 < q < \infty$ and u^ε is bounded in $L^q(\mathbb{R}^{1+d})$, then the set of Λ_k-profiles of u^ε is nonempty. In addition, there is a constant C independent of the family u^ε such that every Λ_k-profile satisfies*

$$\| \mathcal{U}_k \|_{\mathcal{L}_k^q} \;\leq\; C \limsup_{\varepsilon \to 0} \| u^\varepsilon \|_{L^q}\,. \tag{4.3.2}$$

Moreover, if $\mathcal{A} \in \mathcal{L}_k^{q'}(G \times \mathbb{T})$ and

$$v^\varepsilon \sim \tilde{J}_k^\varepsilon(\mathcal{A}) \qquad \text{in} \qquad L^{q'}(\mathbb{R}^{1+d})\,, \tag{4.3.3}$$

then

$$\lim_{\varepsilon \to 0} \int v^{\varepsilon'}\; \overline{u^{\varepsilon'}}\; dt\, dx \;=\; \int \mathcal{A}\, \overline{\mathcal{U}_k}\; dt\, dy\, d\theta\,. \tag{4.3.4}$$

Proof. **a)** Choose G' such that $\jmath_k$ is a diffeomorphism from $G' \subset G$ onto $\Omega' \subset [0,\infty[\times\mathbb{R}^d$. Denote by σ the inverse mapping from Ω' onto G'. Then, for $\mathcal{A} \in \mathcal{P}$ supported in $G' \times \mathbb{T}$,

$$J_k^\varepsilon(\mathcal{A})\,(t,x) \;=\; \frac{1}{\Delta_k(\sigma(t,x))}\,(\mathcal{H}^m\mathcal{A})\,(\sigma(t,x),\varphi(t,x)/\varepsilon)\,, \tag{4.3.5}$$

where $\varphi(t,x) = \psi(\sigma(t,x))$ and m is the value of $m_k(t,y)$ on G'. Assumption 2.2.1 implies that $d\varphi(t,x) \neq 0$ for all $(t,x) \in \Omega'$. Therefore (see [N], [A], [E], [ES], [JMR 2]), there exists a subsequence ε' and $\mathbf{U} \in L^q(\Omega' \times \mathbb{T})$ such that for all $\mathbf{B} \in C_0^\infty(\Omega' \times \mathbb{T})$,

$$\begin{aligned} &\lim_{\varepsilon\to 0}\int_{\Omega'} \mathbf{B}(t,x,\varphi(t,x)/\varepsilon')\,\overline{u^{\varepsilon'}(t,x)}\;dt\,dx \\ &\qquad\qquad = \int \mathbf{B}(t,x,\theta)\,\overline{\mathbf{U}}(t,x,\theta)\;dt\,dx\,d\theta\,, \end{aligned} \tag{4.3.6}$$

In particular, for all $\mathcal{A} \in \mathcal{P}$ supported in G', the integrals

$$\int J_k^{\varepsilon'}(\mathcal{A})\;\overline{u^{\varepsilon'}}\;dt\,dx \tag{4.3.7}$$

converge as $\varepsilon' \to 0$.

b) Since $G\backslash\mathcal{S}_k$ is the countable union of open sets U_α which satisfy the conditions in a), we see that there is a subsequence ε' such that the integrals (4.3.7) converge, for all $\mathcal{A} \in \mathcal{P}_k$. The limit defines a linear form ℓ on $\mathcal{P}_k$. Proposition 4.1.2 applied with the conjugate exponent $q' \in]1,\infty[$, implies that

$$|\,\ell(\mathcal{A})\,| \;\le\; C\,\|\,\mathcal{A}\,\|_{\mathcal{L}_k^{q'}}\;\limsup_{\varepsilon\to 0}\|\,u^\varepsilon\,\|_{L^q}\,. \tag{4.3.8}$$

Since $\mathcal{P}_k$ is dense in $\mathcal{L}_k^{q'}$ this implies that ℓ extends uniquely as a continuous linear form on $\mathcal{L}_k^{q'}$. Hence, there is $\mathcal{U}_k \in \mathcal{L}_k^q$ such that

$$\ell(\mathcal{A}) \;=\; \int \mathcal{A}\,\overline{\mathcal{U}_k}\;dtdyd\theta\,. \tag{4.3.9}$$

This proves (4.3.1), and (4.3.2) follows from (4.3.8).

c) If (4.3.3) holds and $\delta > 0$, introduce the approximations (4.2.4) and use the convergence (4.3.1) for $\mathcal{A}_\delta$. The convergence (4.3.4) follows for the subsequence ε', completing the proof of the proposition.

PROPOSITION 4.3.3. *If $1 < q < \infty$, $\mathcal{U} \in \mathcal{L}_\ell^q$, and $u^\varepsilon \sim \tilde{J}_\ell^\varepsilon(\mathcal{U})$ in $L^q(\mathbb{R}^{1+d})$, then the family u^ε is pure and converges weakly to 0. If $k = \ell$, the Λ_k-profile is equal to $\mathcal{U}$ while if $k \neq \ell$ the Λ_k-profile is equal to 0.*

Proof. **a)** Using the approximations (4.2.4), it suffices to show that for all trigonometric polynomials $\mathcal{A} \in \mathcal{P}_k$ and $\mathcal{B} \in \mathcal{P}_\ell$, one has

$$\lim_{\varepsilon\to 0}\int J_k^\varepsilon(\mathcal{A})\;\overline{J_\ell^\varepsilon(\mathcal{B})}\;dtdx \;=\; \delta_{k,\ell}\int \mathcal{A}\,\overline{\mathcal{B}}\;dtdyd\theta \tag{4.3.10}$$

where $\delta_{k,\ell}$ denotes Kronecker's symbol. One can further reduce to the case where $\mathcal{A}$ and $\mathcal{B}$ are supported in $G' \times \mathbb{T}$ and $G'' \times \mathbb{T}$ respectively, and $\jmath_k$ and $\jmath_\ell$ are diffeomorphisms from G' to Ω' and G'' to Ω'' respectively. Denote by σ' and σ'' the inverse mappings. Then

$$J_k^\varepsilon(\mathcal{A})\,(t,x) \;=\; \mathbf{A}(t,x,\varphi_k(t,x)/\varepsilon)\,, \quad J_k^\varepsilon(\mathcal{B})\,(t,x) \;=\; \mathbf{B}(t,x,\varphi_\ell(t,x)/\varepsilon)\,, \tag{4.3.11}$$

where $\varphi_k := \psi\circ\sigma'$, $\varphi_\ell := \psi\circ\sigma''$,

$$\mathbf{A}(t,x,\theta) \;:=\; \frac{1}{\Delta_k(\sigma'(t,x))}\;(\mathcal{H}^m\mathcal{A})\,(\sigma'(t,x),\theta)\,, \tag{4.3.12}$$

and $\mathbf{B}$ is defined similarly.

b) When $k\neq\ell$, (4.3.10) follows from the convergence

$$\int_{\Omega'\cap\Omega''} \mathbf{A}(t,x,\varphi_k(t,x)/\varepsilon)\;\overline{\mathbf{B}(t,x,\varphi_\ell(t,x)/\varepsilon)}\;dt\,dx \quad\to\quad 0\,. \tag{4.3.13}$$

To prove this convergence, expand $\mathbf{A}$ and $\mathbf{B}$ in Fourier series. Since the mean values of both vanish, it is sufficient to show that for all $\alpha\in\mathbb{Z}\backslash\{0\}$, $\beta\in\mathbb{Z}\backslash\{0\}$ and $a\in C_0^\infty(\Omega'\cap\Omega'')$,

$$\int_{\Omega'\cap\Omega''} a(t,x)\;e^{i\;(\alpha\,\varphi_k(t,x)\,+\,\beta\,\varphi_\ell(t,x)\,)/\varepsilon)}\;dt\,dx \quad\to\quad 0\,. \tag{4.3.14}$$

This convergence follows from the nonresonance Assumption 2.2.5, which implies that $\alpha d\varphi_k+\beta d\varphi_\ell\neq 0$ almost everywhere on $\Omega'\cap\Omega''$.

c) When $k=\ell$, one can assume that $\Omega'=\Omega''$ and $G'=G''$. Expanding $\mathbf{A}$ and $\mathbf{B}$ in Fourier series, and using the property that $d\varphi_k\neq 0$ on Ω', one obtains that

$$\begin{aligned}\int_{\Omega'} \mathbf{A}(t,x,\varphi_k(t,x)/\varepsilon)\;\overline{\mathbf{B}(t,x,\varphi_k(t,x)/\varepsilon)}\;dt\,dx \quad&\to\\ \int_{\Omega'\times\mathbb{T}} \mathbf{A}(t,x,\theta)\;\overline{\mathbf{B}(t,x,\theta)}\;dt\,dx\,d\theta\,.&\end{aligned} \tag{4.3.15}$$

Using (4.3.12) and the fact that Δ_k^2 is the Jacobian of $\jmath_k$, shows that the right hand side is equal to

$$\int_{G'\times\mathbb{T}} (\mathcal{H}^m\mathcal{A})\,(t,y,\theta)\;\overline{(\mathcal{H}^m\mathcal{B})\,(t,y,\theta)}\;dt\,dy\,d\theta\,. \tag{4.3.15}$$

Since $\mathcal{H}$ is unitary, this integral is equal to the right hand side of (4.3.10) and the proof of Proposition 4.3.2 is complete.

PROPOSITION 4.3.4. *Suppose that $1<q\leq 2$, and u^ε is a pure bounded family in L^q with profiles $\mathcal{U}_k$. Then for all $\mathcal{A}\in\mathcal{P}_k(q')$,*

$$\lim_{\varepsilon\to 0}\int I_k^\varepsilon(\mathcal{A})\;\overline{u^\varepsilon}\;dt\,dx \;=\; \int \mathcal{A}\;\overline{\mathcal{U}_k}\;dt\,dy\,d\theta\,. \tag{4.3.16}$$

This is a consequence of Proposition 4.3.1 and the equivalence $I^\varepsilon(\mathcal{A})\sim J^\varepsilon(\mathcal{A})$ in $L^{q'}$ given by Proposition 4.1.3, part *ii)*.

5. The nonlinear interaction operators

In this section we define and study the nonlinear operators $\underline{\mathcal{E}}$ and $\mathcal{E}_k$ which appear in the profile equations (2.5.17). They involve the nonlinear term $f(t,x,u)$ which satisfies Assumption 2.1.2.

5.1. The interaction operators for profiles

Consider $\underline{u} \in L^{p+1}([0,\infty[\times\mathbb{R}^d)$ and for each eigenvalue λ_k, a profile $\mathcal{U}_k(t,y,\theta)$ on $G \times \mathbb{T}$ belonging to the weighted space $\mathcal{L}_k^{p+1}$. Suppose that

$$u^\varepsilon \sim \underline{u} + \sum_k \tilde{J}_k^\varepsilon(\mathcal{U}_k) \quad \text{in } L^{p+1}. \tag{5.1.1}$$

Our goal is to analyse $f(t,x,u^\varepsilon)$.

Suppose that Ω is a regular open set in the sense of Definition 2.2.3. Then $\jmath_k$ is a diffeomorphism from open sets $G_{k,j}$ onto Ω. Denote by $\sigma_{k,j}$ the inverse mappings, and by $m_{k,j}$ the value of $m_k(t,y)$ on $G_{k,j}$. Let Z denote the relevant set of indices (k,j). The formula (2.4.7) introduces profiles $\mathbf{U}_{k,j}$ on $\Omega \times \mathbb{T}$ and, formally, one has on Ω,

$$u^\varepsilon(t,x) \sim \underline{u}(t,x) + \sum_{(k,j)\in Z} \mathbf{U}_{k,j}(t,x,\varphi_{k,j}/\varepsilon), \tag{5.1.2}$$

with $\varphi_{k,j} := \psi \circ \sigma_{k,j}$. Introduce on $\Omega \times (\mathbb{T})^Z$, the function

$$\mathbf{I}(\underline{u},\mathcal{U}_*)\,(t,x,\theta_*) = \underline{u}(t,x) + \sum_{(k,j)\in Z} \mathbf{U}_{k,j}(t,x,\theta_{k,j}), \tag{5.1.3}$$

where $\theta_* := \{\theta_{k,j}\}$ and $\mathcal{U}_* := \{\mathcal{U}_k\}$. Then, formally

$$f(t,x,u^\varepsilon(t,x)) \sim f\big(t,x,\mathbf{I}(\underline{u},\mathcal{U}_*)(t,x,\varphi_*(t,x)/\varepsilon)\big). \tag{5.1.4}$$

As in [JMR 3], we analyze the oscillations of $f(u^\varepsilon)$ by studying the Fourier series of $f(\mathbf{I}(\underline{u},\mathcal{U}_*))$ in θ. Introduce the average over all the angular variables.

$$\underline{\mathcal{E}}(\underline{u},\mathcal{U}_*)\,(t,x) := \int_{(\mathbb{T})^Z} f\big(t,x,\mathbf{I}(\underline{u},\mathcal{U}_*)\,(t,x,\theta_*)\big)\,d\theta_*, \tag{5.1.5}$$

Recall that the total mass of $d\theta$ is equal to 1. The function $\underline{\mathcal{E}}(\underline{u},\mathcal{A}_*)$ is defined on any regular Ω and thus outside $\widetilde{\mathcal{C}} = \bigcup \widetilde{\mathcal{C}}_k$. In particular, it is defined almost everywhere on $[0,\infty[\times\mathbb{R}^d$ as noted in Proposition 2.2.4.

Consider next one index k, and a component $G_{k,j}$ of $\jmath_k^{-1}(\Omega)$. Introduce Z' the complement of $\{k,j\}$ in Z. For $(t,y) \in G_{k,j}$ introduce the average over the angular variables other than the $\{k,j\}^{\text{th}}$

$$\mathcal{F}^{(1)}_{k,j}(\underline{u},\mathcal{U}_*)\,(t,y,\theta) \;:=\; \int_{(\mathbb{T})^{Z'}} f\big(\jmath_k(t,y), \mathbf{I}(\underline{u},\mathcal{U}_*)\,(\jmath_k(t,y),\theta_*)\big)\,d\theta'_*\,. \tag{5.1.6}$$

If θ denotes the omitted $\{k,j\}^{\text{th}}$ variable, then the average in θ of $\mathcal{F}^{(1)}_{k,j}(\underline{u},\mathcal{U}_*)$ is equal to $\underline{\mathcal{E}}(\underline{u},\mathcal{U}_*)(\jmath_k(t,y))$ and thus its oscillating part is

$$\mathcal{F}_{k,j}(\underline{u},\mathcal{U}_*)\,(t,y,\theta) \;:=\; \mathcal{F}^{(1)}_{k,j}(\underline{u},\mathcal{U}_*)\,(t,y,\theta) \;-\; \underline{\mathcal{E}}(\underline{u},\mathcal{U}_*)\,(\jmath_k(t,y))\,. \tag{5.1.7}$$

Finally, lift $\mathcal{F}_{k,j}(\underline{u},\mathcal{U}_*)$ to $G_{k,j}$, introducing

$$\mathcal{E}_k(\underline{u},\mathcal{U}_*)\,(t,y,\theta) \;:=\; \Delta_k(t,y)\,(\mathcal{H}^{-m_{k,j}}\mathcal{F}_{k,j})\,(t,y,\theta)\,. \tag{5.1.8}$$

This defines $\mathcal{E}_k(\underline{u},\mathcal{U}_*)$ on $G_{k,j}$, thus on the disjoint union $\bigcup_j G_{k,j} \;=\; \jmath_k^{-1}(\Omega)$ and therefore outside $\jmath_k^{-1}(\widetilde{\mathcal{C}}_k)$, in particular almost everywhere on $G\times\mathbb{T}$. In the same way , define $\mathcal{F}_k$ and $\mathcal{F}^{(1)}_k$ from their restrictions $\mathcal{F}_{k,j}$ and $\mathcal{F}^{(1)}_{k,j}$

PROPOSITION 5.1.1. *i) If $\underline{u} \in C_0^\infty([0,\infty[\times\mathbb{R}^d)$ and $\mathcal{U}_k \in \mathcal{P}_k$, then $\underline{\mathcal{E}}(\underline{u},\mathcal{U}_*) \in C_0^1([0,\infty[\times\mathbb{R}^d)$ and $\mathcal{E}_k(\underline{u},\mathcal{U}_*) \in C_0^1(G\times\mathbb{T})$.*

ii) The mappings $\underline{\mathcal{E}}$ and $\mathcal{E}_* := \{\mathcal{E}_k\}$ are bounded from $L^{p+1}([0,\infty[\times\mathbb{R}^d) \times \mathcal{L}_*^{p+1}$ to $L^{1+1/p}([0,\infty[\times\mathbb{R}^d)$ and $\mathcal{L}_*^{1+1/p}$ respectively, where $\mathcal{L}_*^q$ denotes the product space $\prod_k \mathcal{L}_k^q$. They are Lipschitz continuous on bounded subsets.

iii) For each k, the support of $\mathcal{E}_k(\underline{u},\mathcal{U}_k)$ is contained in the support of $\mathcal{U}_k$.

Proof. The proof is nalogous to that of Proposition 4.1.2.

a) Suppose that $\underline{u} \in C_0^\infty([0,\infty[\times\mathbb{R}^d)$, and $\mathcal{U}_k \in \mathcal{P}_k$. Since $\mathcal{U}_k$ vanishes on a neighborhood of $\mathcal{S}_k$, all $(\underline{t},\underline{x})$ has a neighborhood Ω' such that $\operatorname{supp}\mathcal{U}_k \cap \jmath_k^{-1}(\Omega')$ is the disjoint union of $G'_{k,j} \subset G$ such that $\jmath_k$ is a diffeomorphism from $G'_{k,j}$ onto Ω'. Thus, only the corresponding variables intervene in the definition of $\mathbf{I}(\underline{u},\mathcal{U}_*)$ implying that $\underline{\mathcal{E}}(\underline{u},\mathcal{U}_*)$ is C^1 with compact support.

A similar argument shows that $\mathcal{E}_k(\underline{u},\mathcal{U}_*)$ belongs to C^1 and vanishes outside a compact, which proves i).

b) For $a \in C_0^\infty(G)$ which vanishes on a neighborhood of $\mathcal{S}_k$, introduce

$$(S_k a)\,(t,x) \;:=\; \sum_{\{y\,;\,q_k(t,y)=x\}} \frac{1}{\Delta_k(t,y)^2}\,a(t,y)\,. \tag{5.1.9}$$

Then, by linearity,

$$\int (S_k a)\,(t,x)\,dt\,dx \;=\; \int a(t,y)\,dt\,dy\,, \tag{5.1.10}$$

since this is true when a is supported in $G' \subset G$ which is diffeomorphic to $\jmath_k(G')$. By density, this extends to $a \in L^1(G)$.

Conversely, consider a nonnegative function $g \in C_0^\infty([0,\infty[\times\mathbb{R}^d)$. The finiteness Assumption 2.2.2 implies that

$$S_k\big(\Delta_k^2\, g\circ\jmath_k\big) \;\le\; m_*\, g\,. \tag{5.1.11}$$

Thus

$$\int \Delta_k(t,y)^2\, g(\jmath_k(t,y))\, dt\, dy \;\le\; m_* \int g(t,x)\, dt\, dx\,. \tag{5.1.12}$$

By density, this extends to $g \in L^1([0,\infty[\times\mathbb{R}^d)$.

c) Consider $\underline{u} \in L^{p+1}$ and $\mathcal{A}_* \in \mathcal{L}_*^{p+1}$. Introduce

$$a_k(t,y) \;:=\; |\Delta_k(t,y)|^{1-p} \int_{\mathbb{T}} |\mathcal{U}_k(t,y,\theta)|^{p+1}\, d\theta\,. \tag{5.1.13}$$

Then

$$\int_G a_k(t,y)\, dt\, dy \;:=\; \|\mathcal{U}_k\|^{p+1}_{\mathcal{L}_k^{p+1}}\,. \tag{5.1.14}$$

Consider $(t,x) \notin \mathcal{C}$ and Ω a regular neighborhood of (t,x). Introduce $\mathbf{I}(\underline{u},\mathcal{A}_*)$ as in (5.1.3). Assumption 2.1.2 implies that

$$|\, f(t,x,\mathbf{I}(\underline{u},\mathcal{U}_*)(t,x,\theta_*))\,|^{1+1/p} \;\le\; C\; |\,\mathbf{I}(\underline{u},\mathcal{U}_*)(t,x,\theta_*))\,|^{p+1}\,. \tag{5.1.15}$$

Integrating in θ, and using the finiteness Assumption 2.2.2, shows that there is a constant C' such that

(5.1.16)

$$|\,\underline{\mathcal{E}}(\underline{u},\mathcal{U}_*)(t,x)\,|^{1+1/p} \;\le\; C'\,\Big(\; |\,\underline{u}(t,x)\,|^{p+1} \;+\; \sum_Z \int |\,\mathbf{U}_{k,j}(t,x,\theta_{k,j})\,|^{p+1}\, d\theta_{k,j}\;\Big)\,.$$

Using the boundedness of the Hilbert transform in $L^{p+1}(\mathbb{T})$ and the notations (5.1.9) (5.1.13), (5.1.16), yields

$$|\,\underline{\mathcal{E}}(\underline{u},\mathcal{U}_*)(t,x)\,|^{1+1/p} \le C'\,\Big(|\underline{u}(t,x)|^{p+1} + \sum_k (S_k a_k)(t,x)\Big) := C' g(t,x)\,. \tag{5.1.17}$$

This estimate holds for $(t,x) \notin \mathcal{C}_k$, and therefore almost everywhere. Integrating and using (5.1.10) (5.1.12), shows that

$$\|\,\underline{\mathcal{E}}(\underline{u},\mathcal{U}_*)\,\|^{1+1/p}_{L^{1+1/p}} \;\le\; C'\,\Big(\; \|\,\underline{u}\,\|^{p+1}_{L^{p+1}} \;+\; \sum_k \;\|\,\mathcal{U}_k\,\|^{p+1}_{\mathcal{L}_k^{p+1}}\;\Big)\,. \tag{5.1.18}$$

d) Similarly, consider $(t,y) \in G$ such that $(t,x) := \jmath_k(t,y) \notin \mathcal{C}_k$. Consider Ω a regular neighborhood of (t,x) and introduce $\mathcal{F}_k$ by (5.1.7). Using (5.1.15), yields

$$\begin{aligned}\int |\,\mathcal{F}_k(t,y,\theta)\,|^{1+1/p}\, d\theta \;&\le\; 2C \int |\,\mathbf{I}(\underline{u},\mathcal{U}_*)\,(\jmath_k(t,y),\theta_*)\,|^{p+1}\, d\theta_* \\ &\le\; C'' g(\jmath_k(t,y))\,.\end{aligned} \tag{5.1.19}$$

Since $\mathcal{H}$ is bounded in $L^{1+1/p}(\mathbb{T})$, the definition (5.1.8) of $\mathcal{E}_k(\underline{u},\mathcal{U}_*)$ shows that

$$\begin{aligned} b_k(t,y) \;&:=\; |\,\Delta_k(t,y)\,|^{1-1/p} \int |\,\mathcal{E}_k(\underline{u},\mathcal{U}_*)(t,y,\theta)\,|^{1+1/p}\, d\theta \\ &\le C''\; \Delta_k(t,y)^2 g(\jmath_k(t,y))\,.. \end{aligned} \tag{5.1.20}$$

This is true for all $(t,y) \notin \jmath_k^{-1}(\mathcal{C}_k)$ hence almost everywhere on G. Integrating over G and using (5.1.12), this proves that

$$\|\,\mathcal{E}_k(\underline{u},\mathcal{U}_*)\,\|^{1+1/p}_{\mathcal{L}_k^{1+1/p}} \;\le\; C'\,\Big(\; \|\,\underline{u}\,\|^{p+1}_{L^{p+1}} \;+\; \sum_k \;\|\,\mathcal{U}_k\,\|^{p+1}_{\mathcal{L}_k^{p+1}}\;\Big)\,. \tag{5.1.21}$$

e) Similarly, using the estimate on the derivatives of f, one shows that

$$\begin{aligned}\| \underline{\mathcal{E}}(\underline{u}, \mathcal{U}_*) - \underline{\mathcal{E}}(\underline{v}, \mathcal{V}_*) \|_{L^{p+1}} + \sum_k \| \mathcal{E}_k(\underline{u}, \mathcal{U}_*) - \mathcal{E}_k(\underline{v}, \mathcal{V}_*) \|_{\mathcal{L}_k^{1+1/p}} \\ \leq C\, K \big(\| \underline{u} - \underline{v} \|_{L^{p+1}} + \sum_k \| \mathcal{U}_k - \mathcal{V}_k \|_{\mathcal{L}_k^{p+1}} \big)\end{aligned} \tag{5.1.22}$$

where

$$K := \| \underline{u} \|_{L^{p+1}}^{p-1} + \| \underline{v} \|_{L^{p+1}}^{p-1} + \sum_k \| \mathcal{U}_k \|_{\mathcal{L}_k^{p+1}}^{p-1} + \| \mathcal{V}_k \|_{\mathcal{L}_k^{p+1}}^{p-1} \tag{5.1.23}$$

f) To prove *iii)* we show that if $\mathcal{U}_k$ vanishes on $G' \subset G$, then $\mathcal{E}_k(\underline{u}, \mathcal{E}_*) = 0$ a.e. on G'. Using Proposition 2.2.4, it is sufficient to show that for all regular open sets Ω, $\mathcal{E}_k(\underline{u}, \mathcal{E}_*) = 0$ a.e. on $G'_{k,j} := G' \cap \jmath_k^{-1}(\Omega)$. This latter set is contained in one component $G_{k,j}$. Since $\mathcal{U}_k = 0$ on G', $\mathbf{I}(\underline{u}, \mathcal{U}_*)$ and $f(t, x, \mathbf{I}(t, x, \theta_*))$ do not depend on the variable $\theta_{k,j}$ for $(t, x) \in \jmath_k(G'_{k,j})$. Formulas (5.1.7) (5.1.8) imply that $\mathcal{E}_k(\underline{u}, \mathcal{U}_*)$ vanishes on $G'_{k,j}$. The proof of Proposition 5.1.1 is complete.

5.2. Profiles of nonlinear functions of oscillations

DEFINITION 5.2.1 *A bounded family h^ε in $L^{1+1/p}([0, \infty[\times\mathbb{R}^d)$ has no propagated oscillations for L if, for all bounded families w^ε in $C^0([0, \infty[\,; L^2(\mathbb{R}^d)) \cap L^{p+1}([0, \infty[\times\mathbb{R}^d)$ such that Lw^ε is bounded in $L^{1+1/p}([0, \infty[\times\mathbb{R}^d)$,*

$$\lim_{\varepsilon \to 0} \int h^\varepsilon \cdot \overline{w^\varepsilon}\, dtdx = 0\,. \tag{5.2.1}$$

The property above is stronger than the fact that $L^{-1}(h^\varepsilon)$ converges to zero in the sense of distributions. Note that if $h^\varepsilon \sim g^\varepsilon$ in $L^{1+1/p}$ and h^ε has no propagated oscillations, then neither does g^ε. Thus the definition extends to asymptotic families, that is to elements of $AL^{1+1/p}$ (with notations of §4.2).

PROPOSITION 5.2.2. *Suppose that $\underline{u} \in L^{p+1}([0, \infty[\times\mathbb{R}^d)$, $\mathcal{U}_k \in \mathcal{L}_k^{p+1}$ and*

$$u^\varepsilon \sim \underline{u} + \sum_k \tilde{J}_k^\varepsilon(\mathcal{U}_k) \quad in \quad L^{p+1}([0, \infty[\times\mathbb{R}^d)\,. \tag{5.2.2}$$

Then $f(u^\varepsilon)$ is a bounded family in $L^{1+1/p}([0, \infty[\times\mathbb{R}^d)$ which is pure in the sense of Definition 4.3.1. Its weak limit is $\underline{\mathcal{E}}(\underline{u}, \mathcal{U}_*)$ and its Λ_k-profiles are $\mathcal{E}_k(\underline{u}, \mathcal{U}_*)$).

Moreover,

$$g^\varepsilon \sim f(u^\varepsilon) - \underline{\mathcal{E}}(\underline{u}, \mathcal{U}_*) - \sum_k \tilde{J}_k^\varepsilon(\mathcal{E}_k(\underline{u}, \mathcal{U}_*)) \quad in \quad L^{1+1/p}([0, \infty[\times\mathbb{R}^d) \tag{5.2.3}$$

has no propagated oscillations for L.

Proof. **a)** Assumption 2.1.2 implies that $f(u^\varepsilon)$ is bounded in $L^{1+1/p}$. It also implies that

$$\| f(u) - f(v) \|_{L^{1+1/p}} \leq C \, (\| u \|_{L^{1+p}} + \| v \|_{L^{1+p}})^{p-1} \, \| u - v \|_{L^{1+p}} .$$

Therefore, changing u^ε by a small term in L^{1+p} induces a small change of $f(u^\varepsilon)$ in $L^{1+1/p}$. By Propositions 5.1.1 and 4.2.2, (5.2.3) defines g^ε as an equivalence class of bounded families in $L^{1+1/p}$. Thus it is sufficient to consider the case where $\underline{u} \in C_0^\infty([0,\infty[\times\mathbb{R}^d)$, $\mathcal{U}_k \in \mathcal{P}_k$, and

$$(5.2.4) \qquad u^\varepsilon = \underline{u} + \sum_k J_k^\varepsilon(\mathcal{U}_k) \, , \quad g^\varepsilon = f(u^\varepsilon) - \underline{\mathcal{E}}(\underline{u}, \mathcal{U}_*) - \sum_k J_k^\varepsilon(\mathcal{E}_k(\underline{u}, \mathcal{U}_*)) \, .$$

Note that Proposition 5.1.1 implies that $\mathcal{E}_k(\underline{u}, \mathcal{U}_*)$ belongs to C^1 and vanishes on a neighborhood of $\mathcal{S}_k$. Therefore the right hand side of (5.2.4) is well defined for each ε. One can further assume that the $\mathcal{U}_k$ are supported in $\jmath_k^{-1}(\Omega)$ where Ω is a regular open set Ω. In this case

$$(5.2.5) \qquad u^\varepsilon(t,x) \;=\; \underline{u}(t,x) \;+\; \sum_Z \; \mathbf{U}_{k,j}(t,x,\varphi_{k,j}/\varepsilon) \, ,$$

where the $\mathbf{U}_{k,j} \in C_0^\infty(\Omega \times \mathbb{T})$ are defined in (2.4.7) and $\varphi_{k,j}$ are the local phases on Ω, introduced in (2.2.9). Then

$$(5.2.6) \qquad f(u^\varepsilon) \;=\; \mathbf{F}(t,x,\varphi_*/\varepsilon) \, ,$$

where

$$(5.2.7) \qquad \mathbf{F}(t,x,\theta_*) \;:=\; f\big(\underline{u}(t,x) + \sum_Z \mathbf{U}_{k,j}(t,x,\theta_{k,j})\big) \;\in C_0^\infty(\Omega \times \mathbb{T}^Z) \, .$$

Consider the rapidly convergent Fourier expansion

$$(5.2.8) \qquad \mathbf{F}(t,x,\theta_*) \;=\; \sum_{\alpha_* \in (\mathbb{Z})^Z} \; c_{\alpha_*}(t,x) \; e^{i\,\alpha_* \cdot \theta_*} \, .$$

Then, the definition of $\underline{\mathcal{E}}$ and $\mathcal{E}_k$ is made so that

$$(5.2.9) \qquad g^\varepsilon \;=\; \sum_{\alpha_* \in R} \; c_{\alpha_*}(t,x) \; e^{i\,\alpha_* \cdot \theta_*} \, ,$$

with R the set of those $\alpha_* \in \mathbb{Z}^Z$ such that at least two components $\alpha_{k,j}$ do not vanish.

b) To compute the profiles, it suffices to show that, for all $\alpha_* \in R$ and $c \in C_0^\infty(\Omega)$, the sequence

$$(5.2.10) \qquad \ell^\varepsilon(t,x) \;:=\; c(t,x) \; e^{i\;\alpha_*\,\varphi_*(t,x)/\varepsilon}$$

converges weakly to 0, and that all its profiles vanish.

Assumption 2.2.5 implies that the phase $\varphi := \alpha_* \varphi_*$ satisfies $d\varphi \neq 0$ almost everywhere. Therefore ℓ^ε converges weakly to 0.

Consider $\mathcal{A} \in \mathcal{P}_k$. On Ω, $J_k^\varepsilon(\mathcal{A})$ is of the form

$$(5.2.11) \quad J_k^\varepsilon(\mathcal{A})(t,x) = \sum_j \mathbf{A}_j(t,x,\varphi_{k,j}(t,x)/\varepsilon) \;= \sum_{j,n} a_{j,n}(t,x) \, e^{i\;n\;\varphi_{k,j}(t,x)/\varepsilon} \, ,$$

where n runs over a finite subset of $\mathbb{Z}$. The phase $\alpha_*\varphi_* + n\varphi_{k,j}$ is either of the form $\gamma\varphi_{k',j'}$ with $\gamma \neq 0$, or of the form $\alpha'_*\varphi_*$ with $\alpha'_* \in R$. In the first case, the differential of the phase does not vanish by Assumption 2.2.1. In the second case, it does not vanish almost everywhere by Assumption 2.2.5. This implies that

$$\lim_{\varepsilon\to 0} \int \ell^\varepsilon(t,x)\ J_k^\varepsilon(\mathcal{A})(t,x)\ dtdx \ = \ 0\,, \tag{5.2.12}$$

showing that ℓ^ε is k-pure with Λ_k-profile equal to 0.

c) It remains to show that ℓ^ε defined in (5.2.10) has no propagated oscillations for L. Assumption 2.2.5 implies that $\det L(d\varphi) \neq 0$ almost everywhere on Ω. Thus the Lebesgue measure of the set of points (t,x) where $|\det L(d\varphi(t,x))| \leq \delta\}$ tends to 0 as δ tends to 0, implying that the $L^{1+1/p}$ norm of ℓ^ε on this set converges to 0 as $\delta \to 0$, uniformly with respect to ε. Therefore, it is sufficient to prove the convergence (5.2.1) for $\ell^\varepsilon = c\,e^{i\varphi/\varepsilon}$, when c is compactly supported in the open set where $\det L(d\varphi) \neq 0$.

In this case,

$$b \ := \ L(i\,d\varphi)^{-1}c \ \in \ C_0^\infty(\Omega) \tag{5.2.13}$$

and

$$c\,e^{i\varphi/\varepsilon} \ = \ \varepsilon\ L(b\,e^{i\,\varphi/\varepsilon}) \ - \ \varepsilon\ (Lb)\ e^{i\varphi/\varepsilon}\,. \tag{5.2.14}$$

Suppose that w^ε is a bounded family in L^{1+p} such that Lw^ε is bounded in $L^{1+1/p}$. Then using (5.2.13) and (5.2.14), integration by parts show that

$$\int c\ e^{i\varphi/\varepsilon}\cdot \overline{w^\varepsilon}\ dtdx \ = \ O(\varepsilon)\,. \tag{5.2.15}$$

This finishes the proof for ℓ^ε.

Recall that the spectral projector P_k is defined in Proposition 2.5.2.

PROPOSITION 5.2.3. *If $\mathcal{A} \in \mathcal{L}_k^{1+1/p}$ and $P_k\mathcal{A} = 0$, then $\tilde{J}_k^\varepsilon(\mathcal{A})$ has no propagated oscillations for L.*

Proof. Using the remark following Definition 5.2.1 and Proposition 4.2.2, it suffices to prove the result when $\mathcal{A} \in \mathcal{P}_k$ and is a monomial : $\mathcal{A}(t,y,\theta) = a(t,y)\ e^{in\theta}$. Then $J_k^\varepsilon(\mathcal{A})$ is a sum of terms of the form

$$c_{k,j}(t,x)\ e^{i\,n\,\varphi_{k,j}/\varepsilon}\,, \tag{5.2.16}$$

where $c_{k,j}(t,x) := \Delta_k^{-1}(t,\rho_{k,j}(t,x))\,a((t,\rho_{k,j}(t,x))$. Let $P_{k,j}(t,x) := P_k(\rho_{k,j}(t,x))$ denote the orthogonal projector on the kernel of $L(d\varphi_{k,j})$. It remains to show that (5.2.1) is satisfied when $h^\varepsilon = c\,e^{i\,\varphi/\varepsilon}$ with $\varphi := \varphi_{k,j}$ and $c \in C_0^\infty(\Omega)$ satisfies $P_{k,j}c = 0$. In such case, there exists $b \in C_0^\infty(\Omega)$ such that $L(id\varphi)b = c$. Then, (5.2.14) holds, implying (5.2.15), and the proof of Proposition 5.2.3 is complete.

COROLLARY 5.2.4. *For all $\mathcal{F}_k \in \mathcal{L}_k^{1+1/p}$, $\tilde{J}_k^\varepsilon((I-P_k)\mathcal{F}_k)$ has no propagated oscillations for L.*

6. Proof of asymptotics

In this section we prove Theorems 2.5.1 and 2.5.4 and show that the profile equations (2.5.16) are satisfied. The strategy is the following. We extract from both families u^ε and $f^\varepsilon := f(u^\varepsilon)$ pure subsequences, in the sense of §4. Equations are derived for the weak limits $\underline{u}$ and $\underline{f}$ and their profiles $\mathcal{U}_k$ and $\mathcal{F}_k$. Next, we introduce an appropriate $v^\varepsilon \sim \underline{u} + \sum \tilde{J}_k^\varepsilon(\mathcal{U}_k)$, such that $Lv^\varepsilon \sim \underline{f} + \sum \tilde{J}_k^\varepsilon(P_k\mathcal{F}_k)$. Using the dissipativity of the equation, we prove that the subsequence $u^\varepsilon - v^\varepsilon$ converges strongly to 0. This property is used to identify the weak limits $\underline{f}$ and the profiles $\mathcal{F}_k$ as functions of $\underline{u}$ and $\mathcal{U}_k$ and to deduce that $\underline{u}$ and $\mathcal{U}_k$ satisfy the profile equations. Finally, we show that the solution of the profile equation is unique, implying that no extraction of subsequences was necessary and that $u^\varepsilon \sim \underline{u} + \sum \tilde{J}_k^\varepsilon(\mathcal{U}_k)$ as claimed.

6.1. Weak convergence

Consider the family of solutions u^ε of (2.1.1) with Cauchy data (2.5.1). Then, u^ε is bounded in $C^0([0,\infty[\,; L^2(\mathbb{R}^d))$ and in $L^{p+1}([0,\infty[\times\mathbb{R}^d)$. This implies that $f^\varepsilon := f(t,x,u^\varepsilon)$ is bounded in $L^{1+1/p}([0,\infty[\times\mathbb{R}^d)$ By Proposition 4.3.2 we can extract pure subsequences which, for simplicity, we still call u^ε and f^ε.

Recall that $P_k(y)$ denotes the orthogonal projector on the kernel of $\lambda_k(\xi)Id - \sum \xi_j A_j$ for $\xi = d\psi(y)$ and $\mathcal{S}_k^+(p+1)$ is the set of supercritical points introduced in Definition 2.3.4.

PROPOSITION 6.1.1. *The Λ_k-profiles $\mathcal{U}_k$ of $\{u^\varepsilon\}$, $\mathcal{F}_k$ of $\{f^\varepsilon\}$ satisfy $\mathcal{U}_k \in \mathcal{L}_k^{p+1}$, $\mathcal{F}_k \in \mathcal{L}_k^{1+1/p}$, and the polarization condition*

$$P_k\,\mathcal{U}_k = \mathcal{U}_k\,. \tag{6.1.1}$$

Moreover, they satisfy on $(G\backslash\mathcal{S}_k^+(p+1)) \times \mathbb{T}$

$$\partial_t\mathcal{U}_k \;+\; P_k\,\mathcal{F}_k \;=\; 0\,, \qquad \mathcal{U}_k|_{t=0} \;=\; P_k\,\mathcal{U}_0\,. \tag{6.1.2}$$

The weak limits satisfy $\underline{u} \in L^{p+1}([0,\infty[\times\mathbb{R}^d)$, $\underline{f} \in L^{1+1/p}([0,\infty[\times\mathbb{R}^d)$ and

$$L\underline{u} \;+\; \underline{f} \;=\; 0 \quad \text{on } [0,\infty[\times\mathbb{R}^d\,, \qquad \underline{u}|_{t=0} \;=\; \underline{u}_0\,. \tag{6.1.3}$$

Proof. **a)** Let $\mathcal{A}$ be a trigonometric polynomial. For ε fixed, $I_k^\varepsilon(\mathcal{A})$ is rapidly decreasing as $|x| \to \infty$ (see parts b) and c) in the proof of Theorem 3.5.5) and is compactly supported in t. Let $a_0^\varepsilon := \mathcal{A}(0, y, \psi(y)/\varepsilon)$. Then, since L is skew adjoint, one has

$$(6.1.4) \qquad (Lu^\varepsilon, I_k^\varepsilon(\mathcal{A})) \;=\; -\,(u^\varepsilon, LI_k^\varepsilon(\mathcal{A})) \;-\; (u_0^\varepsilon, a_0^\varepsilon)\,,$$

where (,) denotes scalar product in L^2.

b) One has (see e.g. [Hö 2], Theorem 7.7.7)

$$(6.1.5) \qquad L(I_k^\varepsilon(\mathcal{A})) \;\sim\; \frac{1}{\varepsilon}\, I_k^\varepsilon(\mathcal{B}_{-1}) \;+\; I_k^\varepsilon(\mathcal{B}_0) \;+\; O(\varepsilon) \quad \text{in } L^2 \cap L^\infty([0,\infty[\times\mathbb{R}^d)\,,$$

with

$$(6.1..6) \qquad \mathcal{B}_{-1} \;:=\; i\,\big(\lambda_k(d\psi)\,Id \;-\; A(d\psi)\big)\mathcal{A}\;, \qquad \mathcal{B}_0 \;:=\; \mathcal{D}_k\,\mathcal{A}\,,$$

$$(6.1.7) \qquad \mathcal{D}_k \;:=\; \frac{\partial}{\partial t} \;+\; \sum_j \big(\frac{\partial \lambda_k}{\partial \xi_j}(d\psi)\,Id - A_j\big)\,\frac{\partial}{\partial y_j} \;+\; \gamma\,.$$

$$(6.1.8) \qquad \gamma \;:=\; \frac{1}{2}\sum_l\sum_j \frac{\partial^2 \lambda_k}{\partial \xi_j\,\partial \xi_l}(d\psi)\,\frac{\partial^2 \psi}{\partial y_j\,\partial y_l}$$

Differentiating twice the identity $(\lambda_k(\xi)Id - A(\xi))\Pi_k(\xi) = 0$, where $\Pi_k(\xi)$ is the spectral projector of $A(\xi)$, and multiplying the result on the left by $\Pi_k(\xi)$, one proves the fundamental identity

$$(6.1.9) \qquad P_k\,\mathcal{D}_k\,(P_k\,\mathcal{A}) \;:=\; \partial_t\,P_k\mathcal{A}\,.$$

c) Suppose that $\mathcal{B} \in \mathcal{P}_k$ satisfies $P_k\mathcal{B} = 0$. Then there is a unique $\mathcal{A} \in \mathcal{P}_k$ such that

$$(6.1.10) \qquad \mathcal{B} \;:=\; i\,\big(\lambda_k(d\psi)\,Id \;-\; A(d\psi)\big)\mathcal{A}\;, \qquad P_k\mathcal{A} = 0\,.$$

Therefore, (6.1.5) implies that $\varepsilon L I_k^\varepsilon(\mathcal{A})) \sim I_k^\varepsilon(\mathcal{B})$ in L^2. Since $\mathcal{A}$ vanishes near $\mathcal{S}_k$, $I_k^\varepsilon(\mathcal{A})$ is bounded in L^q for all q. Thus (6.1.4) implies that for all such $\mathcal{B}$,

$$(6.1.11) \qquad \int \mathcal{U}_k \cdot \overline{\mathcal{B}}\; dtdyd\theta \;=\; 0\,,$$

which proves (6.1.1).

d) For $\mathcal{A} \in \mathcal{P}_k(p+1)$ (see Definition 4.1.1) such that $P_k\mathcal{A} = \mathcal{A}$, (6.1.5) and (6.1.9) imply that $LI_k^\varepsilon(\mathcal{A}) \sim I^\varepsilon(\partial_t\mathcal{A})$ in L^2. Since u^ε is pure in L^2, this implies that

$$(6.1.12) \qquad (u^\varepsilon,\, LI_k^\varepsilon(\mathcal{A})) \;\to\; \int \mathcal{U}_k \cdot \overline{\partial_t\mathcal{A}}\; dtdyd\theta\,.$$

Since the subsequence f^ε is pure in $L^{1+1/p}$, Proposition 4.3.4 implies that

$$(6.1.13) \qquad (f^\varepsilon,\, I_k^\varepsilon(\mathcal{A})) \;\to\; \int \mathcal{F}_k \cdot \overline{\mathcal{A}}\; dtdyd\theta\,.$$

Finally, the Cauchy data (2.5.1) u_0^ε is pure in L^2 and we can pass to the limit in (6.1.4) to find

$$\int \mathcal{F}_k \cdot \overline{\mathcal{A}}\, dtdyd\theta \;=\; \int \mathcal{U}_k \cdot \overline{\partial_t \mathcal{A}}\, dtdyd\theta \;+\; \int \mathcal{U}_0 \cdot \overline{\mathcal{A}}|_{t=0}\, dyd\theta\,. \tag{6.1.14}$$

For $\mathcal{B} \in \mathcal{P}_k(p+1)$, one can apply (6.1.14) to $\mathcal{A} = P_k\mathcal{B}$. Using (6.1.1), we see that

$$\int P_k\, \mathcal{F}_k \cdot \overline{\mathcal{B}}\, dtdyd\theta \;=\; \int \mathcal{U}_k \cdot \overline{\partial_t \mathcal{B}}\, dtdyd\theta \;+\; \int P_k\, \mathcal{U}_0 \cdot \overline{\mathcal{B}}|_{t=0}\, dyd\theta\,. \tag{6.1.15}$$

This proves (6.1.2).

e) The equation (6.1.3) for the weak limits is immediate.

6.2. Construction of approximate solutions

The next result is fundamental in the understanding of equations (6.1.2).

PROPOSITION 6.2.1. *Suppose that* $\mathcal{U} \in \mathcal{L}_k^{p+1}$ *and* $\mathcal{F} \in \mathcal{L}_k^{1+1/p}$ *satisfy* $\partial_t\mathcal{U} = \mathcal{F}$ *on* $(G\backslash\mathcal{S}_k^+(p+1)) \times \mathbb{T}$ *and* $\mathcal{U}|_{t=0} \in L^2(\omega \times \mathbb{T})$. *Then there exist trigonometric polynomials* $\mathcal{U}_n \in \mathcal{P}_k(p+1)$ *and* $\mathcal{F}_n \in \mathcal{P}_k$ *such that*

i) $\partial_t\, \mathcal{U}_n \;=\; \mathcal{F}_n$ *on* $G \times \mathbb{T}$

ii) $\mathcal{U}_n \to \mathcal{U}$ *strongly in* $\mathcal{L}_k^{p+1}$,

iii) $\mathcal{F}_n \to \mathcal{F}$ *strongly in* $\mathcal{L}_k^{1+1/p}$,

iv) $\mathcal{U}_n|_{t=0} \to \mathcal{U}|_{t=0}$ *strongly in* $L^2(\omega \times (\mathbb{R}/2\pi\mathbb{Z}))$.

When $\mathcal{U}$ *satisfies* $P_k\mathcal{U} = \mathcal{U}$, *one can choose the approximations* $\mathcal{U}_n$ *such that* $P_k\mathcal{U}_n = \mathcal{U}_n$.

Proof. **a)** Multiplying $\mathcal{U}$ by a function $\chi(y)$ affects neither the assumptions nor the result. Since $\omega\backslash\omega_0$ has measure by Assumption 2.3.3, a cutoff of this type yields an approximation with support in $[0,\infty[\times\omega_0$. In addition the proof of Proposition 6.2.1 is trivial when $\mathcal{U}$ is supported in a tube $[T_1, T_2] \times \omega'$ which does not intersect $\mathcal{S}_k$. Therefore one can assume that $\mathcal{U}(t,y)$ and $\mathcal{F}(t,y)$ are supported in a tube $]a,b[\times\omega'$ where $\omega' \subset \omega_0$ is a small open ball such that $g_k'(y)$ has a unique positive eigenvalue, $1/r(y)$, of constant multiplicity μ, such that $a < a' \le r(y) \le b' < b$ for all $y \in \omega'$. Then

$$\Delta_k(t,y) \;\approx\; |t - r(y)|^{\mu/2}\,. \tag{6.2.1}$$

There are two cases. First $\mathcal{S}_k$ intersects $]a,b[\times\omega'$ at supercritical points, which corresponds to $\alpha := (p-1)\mu/2 \ge 1$, and the equation $\partial_t\mathcal{U} = \mathcal{F}$ holds outside $\mathcal{S}_k$. Second the intersection is subcritical, that is $\alpha < 1$ and the equation holds throughout $]a,b[\times\omega'$.

Changing coordinates $s := t - r(y)$, and setting $z := (y,\theta)$ we are reduced to the situation where $u(s,z)$ and $f(s,z)$ are compactly supported functions on $V :=]-1,+1[\times V'$, such that

$$\int_V |s|^{-\alpha}\, |u|^{p+1}\, dsdz \;<+\infty\,, \qquad \int_V |s|^{\alpha/p}\, |f|^{1+1/p}\, dsdz \;<+\infty\;, \tag{6.2.2}$$

and

$$\partial_s u = f \quad \text{on } V\,, \qquad \text{if } \alpha < 1\,, \tag{6.2.3}$$

$$\partial_s u = f \quad \text{on } V \cap \{s \neq 0\}\,, \qquad \text{if } \alpha \geq 1\,. \tag{6.2.4}$$

Mollifying in z, we can approximate u and f by trigonometric polynomials with coefficients infinitely smooth in y. It remains to smooth the functions with respect to the variable s.

b) When $\alpha < 1$, f is integrable since Hölder's inequality yields

$$\int_V |f| \leq \Big(\int_V |s|^{\alpha/p} |f|^{1+1/p}\Big)^{p/(p+1)} \Big(\int_V |s|^{-\alpha}\Big)^{1/(p+1)} < +\infty\,. \tag{6.2.5}$$

Thus, u is continuous. Mollifying in s yields approximations which converge uniformly and therefore in the weighted spaces since $|s|^{-\alpha}$ is integrable.

c) Suppose $\alpha \geq 1$. Let $\chi(s)$ be equal to 1 for $|s| \geq 1$ and equal to 0 for $|s| \leq 1/2$. For $\delta > 0$, let $\chi_\delta(s) := \chi(s/\delta)$, $u_\delta := \chi_\delta u$ and $f_\delta := \chi_\delta f$. Then

$$\partial_s u_\delta = f_\delta \; + g_\delta\,, \quad \text{with } g_\delta := \frac{1}{\delta}\,\chi'(\frac{s}{\delta})\,u\,. \tag{6.2.6}$$

Thus, with $\beta := \alpha/p$,

$$\begin{aligned} &\int_V |s|^\beta\, |g_\delta|^{1+1/p} \leq C\,\delta^{\beta-1-1/p} \int_{V_\delta} |u|^{1+1/p} \leq \\ &C\,\delta^{\beta-1+1-2/p}\Big(\int_{V_\delta} |u|^{p+1}\Big)^{1/p} \leq C\,\delta^{2(\alpha-1)/p}\Big(\int_{V_\delta} |s|^{-\alpha}|u|^{p+1}\Big)^{1/p}\,, \end{aligned} \tag{6.2.7}$$

where V_δ is the set of (s, z) in V such that $\delta/2 \leq |s|\ \delta$. Since $\alpha \geq 1$, the right hand side of (6.2.7) tends to zero as δ tends to zero. This constructs approximations which vanish on a neighborhood of $s = 0$. Smoothing such functions away from $s = 0$ is easy.

d) If $\mathcal{U}$ satisfies the polarization condition $P_k\mathcal{U} = \mathcal{U}$, approximate $\mathcal{U}$ by $\mathcal{U}_n$ as above, and then replace $\mathcal{U}_n$ by $P_k\mathcal{U}_n$.

COROLLARY 6.2.2. *Suppose that $\mathcal{U}$ and $\mathcal{F}$ satisfy the assumptions of Proposition 6.2.1. Then $\mathcal{U} \in C^0([0,\infty[\,; L^2(\omega \times \mathbb{T}))$ and*

$$\|\,\mathcal{U}(T)\,\|^2_{L^2(\omega\times\mathbb{T})} = \|\,\mathcal{U}(0)\,\|^2_{L^2(\omega\times\mathbb{T})} + 2\,\mathrm{Re}\Big(\int_{[0,T]\times\omega\times\mathbb{T}} \mathcal{F}\,\overline{\mathcal{U}}\;dt\,dyd\theta\Big)\,. \tag{6.2.8}$$

Moreover, the sequence $\mathcal{U}_n$ given by Proposition 6.1.1 converges strongly to $\mathcal{U}$ in $C^0([0,\infty[\,; L^2(\omega \times \mathbb{T}))$.

Proof. The estimate (6.2.8) is clear for smooth $\mathcal{U}$ and $\mathcal{F}$. Applying it for $\mathcal{U}_n - \mathcal{U}_{n'}$, implies that $\mathcal{U}_n$ is a Cauchy sequence in $C^0([0,\infty[\,; L^2(\omega\times\mathbb{T}))$. Therefore, $\mathcal{U}$ belongs to this space and the convergence holds in $C^0([0,\infty[\,; L^2(\omega \times \mathbb{T}))$. Passing to the limit this implies that $\mathcal{U}$ also satisfies (6.2.8).

PROPOSITION 6.2.3. *Suppose that*

$$u \in L^{p+1}([0,\infty[\times\mathbb{R}^d)\,, \quad Lu \in L^{1+1/p}([0,\infty[\times\mathbb{R}^d)\,, \quad u|_{t=0} \in L^2(\mathbb{R}^d)\,.$$

Then $u \in C^0([0,\infty[\,;L^2(\mathbb{R}^d))$ and there are C^∞ functions with compact support u_n such that

i) $u_n \to u$ strongly in $L^{p+1}([0,\infty[\times\mathbb{R}^d)$ and in $C^0([0,\infty[\,;L^2(\mathbb{R}^d))$,
ii) $Lu_n \to Lu$ strongly in $L^{1+1/p}([0,\infty[\times\mathbb{R}^d)$.
iii) $u_n|_{t=0} \to u|_{t=0}$ strongly in $L^2(\mathbb{R}^d)$.

Moreover

$$(6.2.9) \qquad 2\,\mathrm{Re}\int_{[0,T]\times\mathbb{R}^d} Lu\cdot\overline{u}\;dt\,dx \;=\; \|u(T)\|^2_{L^2} \;-\; \|u(0)\|^2_{L^2}\,.$$

Proof. The approximations are constructed by standard mollification. The usual energy identity (6.2.9) is satisfied for smooth u_n. Taking limits, it implies that $u \in C^0([0,T]\,;L^2(\mathbb{R}^d))$, and satisfies (6.2.9).

Our goal is to prove that $u^\varepsilon \sim \underline{u} + \sum \tilde{J}^\varepsilon_k(\mathcal{U}_k)$. The strategy is to introduce

$$(6.2.10) \qquad v^\varepsilon \;\sim\; \underline{u} \;+\; \sum \tilde{J}^\varepsilon_k(\mathcal{U}_k) \quad \text{in} \quad L^{p+1}([0,T]\times\mathbb{R}^d)\,,$$

and to prove that $\|u^\varepsilon - v^\varepsilon\|_{L^{p+1}} \to 0$. To do so, we use the energy estimates and the monotonicity of the equation. In particular we use $Lv^\varepsilon \in L^{1+1/p}$. A difficulty is that (6.2.10) defines v^ε up to a sequence which converges strongly to zero in L^{p+1}, and Lv^ε may not belong to $L^{1+1/p}$. The next proposition shows that, knowing (6.1.2) (6.1.3), one can choose a representative v^ε such that Lv^ε is bounded in $L^{1+1/p}$.

PROPOSITION 6.2.4. *Suppose that $\underline{u}$, $\underline{f}$, $\mathcal{U}_k$ and $\mathcal{F}_k$ are as in Proposition 6.1.1. Then, for all $T > 0$, there exist a bounded family v^ε in $C^0([0,T]\,;L^2(\mathbb{R}^d)) \cap L^{p+1}([0,T]\times\mathbb{R}^d)$, such that Lv^ε is bounded in $L^{1+1/p}([0,T]\times\mathbb{R}^d)$ and satisfies (6.2.10) and*

$$(6.2.11) \qquad Lv^\varepsilon \;\sim\; \underline{f} \;+\; \sum \tilde{J}^\varepsilon_k(P_k\mathcal{F}_k) \quad \text{in} \quad L^{1+1/p}([0,T]\times\mathbb{R}^d)\,,$$

$$(6.2.12) \qquad v^\varepsilon|_{t=0} \;\sim\; \underline{u}_0 \;+\; \mathcal{U}_0(y,\psi(y)/\varepsilon) \quad \text{in} \quad L^2(\mathbb{R}^d)\,.$$

Proof. a) Choose

$$\begin{aligned} w^\varepsilon_k \;&\sim\; \tilde{J}^\varepsilon_k(\mathcal{U}_k) \quad \text{in} \quad L^{p+1}([0,T]\times\mathbb{R}^d)\,,\\ h^\varepsilon_k \;&\sim\; \tilde{J}^\varepsilon_k(P_k\mathcal{F}_k) \quad \text{in} \quad L^{1+1/p}([0,T]\times\mathbb{R}^d)\,,\\ w^\varepsilon_0 \;&\sim\; \tilde{J}^\varepsilon_0(\mathcal{U}_0) \quad \text{in} \quad L^2(\mathbb{R}^d)\,, \end{aligned}$$

We look for v^ε, as

$$(6.2.13) \qquad v^\varepsilon \;:=\; \underline{u} \;+\; \sum v^\varepsilon_k\,.$$

Since $\underline{u} \in L^{p+1}$, $\underline{f} \in L^{1+1/p}$ and $L\underline{u} = \underline{f}$, Proposition 6.2.3 shows that $u \in C^0([0,\infty[\,; L^2(\mathbb{R}^d))$. The same proposition shows that it is sufficient to construct v_k^ε such that

$$v_k^\varepsilon \sim \tilde{J}_k^\varepsilon(\mathcal{U}_k) \quad \text{in} \quad L^{p+1}([0,T] \times \mathbb{R}^d)\,, \tag{6.2.14}$$

$$Lv^\varepsilon \sim \tilde{J}_k^\varepsilon(P_k\mathcal{F}_k) \quad \text{in} \quad L^{1+1/p}([0,T] \times \mathbb{R}^d)\,, \tag{6.2.15}$$

and

$$v_k^\varepsilon|_{t=0} \sim P_k\mathcal{U}_0(y, \psi(y)/\varepsilon) \quad \text{in } L^2(\mathbb{R}^d)\,. \tag{6.2.16}$$

b) Fix $\delta > 0$ and $T > 0$. Propositions 6.1.1 and 6.2.1 imply that there are $\mathcal{V}_k \in \mathcal{P}_k(p+1)$ and $\mathcal{G}_k \in \mathcal{P}_k$, such that

$$\|\, \mathcal{V}_k - \mathcal{U}_k \,\|_{L^{1+p}_{w,k}(G\times\mathbb{T})} \le \delta\,, \qquad \|\, \mathcal{V}_k|_{t=0} - P_k\mathcal{U}_0 \,\|_{L^2(\omega\times\mathbb{T})} \le \delta\,, \tag{6.2.17}$$

$$\|\, \mathcal{G}_k - P_k\mathcal{F}_k \,\|_{\mathcal{L}_k^{1+1/p}(G\times\mathbb{T})} \le \delta\,, \tag{6.2.18}$$

and

$$P_k\mathcal{V}_k = \mathcal{V}_k\,, \quad \partial_t\mathcal{V}_k = \mathcal{G}_k\,. \tag{6.2.19}$$

The identities (6.1.5) (6.1.6) and (6.1.9) imply that

$$LI_k^\varepsilon(\mathcal{V}_k) - I_k^\varepsilon(\mathcal{G}_k) = O(\varepsilon) \quad \text{in} \quad L^2 \cap L^\infty([0,T] \times \mathbb{R}^d)\,. \tag{6.2.20}$$

Introduce

$$v_k^{\varepsilon,\delta} := I_k^\varepsilon(\mathcal{V}_k)\,. \tag{6.2.21}$$

Since $\mathcal{V}_k \in \mathcal{P}_k(p+1)$, part *ii)* of Proposition 4.1.3 implies that $I_k^\varepsilon(\mathcal{V}_k) \sim J_k^\varepsilon(\mathcal{V}_k)$ in L^{1+p}. Moreover, (6.2.17) and Proposition 4.2.2 imply that $J_k^\varepsilon(\mathcal{V}_k) - \tilde{J}_k^\varepsilon(\mathcal{U}_k)$ is $O(\delta)$ in AL^{p+1}. Therefore, there is $\varepsilon(\delta) > 0$, such that for all $\varepsilon < \varepsilon(\delta)$,

$$\|\, v_k^{\varepsilon,\delta} - w_k^\varepsilon \,\|_{L^{1+p}} \le C\,\delta\,. \tag{6.2.22}$$

Similarly, since $\mathcal{G}_k \in \mathcal{P}_k$, $\mathcal{G}_k$ vanishes near $\mathcal{S}_k$ and $I_k^\varepsilon(\mathcal{G}_k) \sim J_k^\varepsilon(\mathcal{G}_k)$ in L^q for all $q \in [1,\infty]$. With (6.2.20), (6.2.18) and Proposition 4.2.2, this implies that one can choose $\varepsilon(\delta,T) > 0$, such that for all $\varepsilon < \varepsilon(\delta,T)$,

$$\|\, Lv_k^{\varepsilon,\delta} - h_k^\varepsilon \,\|_{L^{1+1/p}} \le \delta\,. \tag{6.2.23}$$

Finally, (6.2.17) together with the initial condition for $\mathcal{U}_k$ in (6.1.2) show that, for $\varepsilon < \varepsilon(\delta,T)$

$$\|\, v_k^{\varepsilon,\delta}|_{t=0} - P_k\, w_0^\varepsilon \,\|_{L^2(\mathbb{R}^d)} \le \delta\,. \tag{6.2.24}$$

One can assume that $\varepsilon(\delta,T)$ decreases as δ decreases.

c) There is a positive function $\delta(\varepsilon)$ such that $\varepsilon < \varepsilon(\delta(\varepsilon))$ and $\delta(\varepsilon) \to 0$ as $\varepsilon \to 0$. Therefore, (6.2.22) (6.2.23) (6.2.24) imply that $v_k^\varepsilon := v_k^{\varepsilon,\delta(\varepsilon)}$ satisfies (6.2.14) (6.2.15) and (6.2.16), and Proposition 6.2.4 is proved.

6.3. Strong convergence.

PROPOSITION 6.3.1. *If u^ε, $\underline{u}$, $\mathcal{U}_k$, $\underline{f}$, $\mathcal{F}_k$, T and v^ε are as in Proposition 6.2.4, then $u^\varepsilon - v^\varepsilon$ converges strongly to zero in $C^0([0,T]\,;L^2(\mathbb{R}^d))$ and in $L^{1+p}([0,T]\times\mathbb{R}^d)$. In particular,*

$$u^\varepsilon \sim \underline{u} + \sum \tilde{J}_k^\varepsilon(\mathcal{U}_k) \qquad in \quad L^{p+1}([0,T]\times\mathbb{R}^d)\,, \tag{6.3.1}$$

Proof. **a)** u^ε and v^ε are bounded in L^{p+1} and Lu^ε and Lv^ε are bounded in $L^{1+1/p}$. Therefore, Proposition 6.2.3 implies that

$$\| u^\varepsilon(t) - v^\varepsilon(t) \|_{L^2}^2 \; - \; 2\,\mathrm{Re} \int_{[0,t]\times\mathbb{R}^d} L(u^\varepsilon - v^\varepsilon)\cdot(\overline{v^\varepsilon} - \overline{u^\varepsilon})\; dt\, dx \; = \; \| u^\varepsilon(0) - v^\varepsilon(0) \|_{L^2}^2 \,. \tag{6.3.2}$$

Introduce

$$h^\varepsilon \; := \; L(v^\varepsilon - u^\varepsilon) \; + \; f(v^\varepsilon) - f(u^\varepsilon) \; = \; Lv^\varepsilon \; + \; f(v^\varepsilon)\,. \tag{6.3.3}$$

Substituting $L(u^\varepsilon - v^\varepsilon) = -\,(f(u^\varepsilon) - f(v^\varepsilon)) + h^\varepsilon$ in (6.3.2) and using Assumption 2.1.2, we obtain that

$$\| u^\varepsilon(t) - v^\varepsilon(t) \|_{L^2}^2 \; + \; c\, \| u^\varepsilon - v^\varepsilon \|_{L^{1+p}([0,t]\times\mathbb{R}^d)}^{1+p} \; \le \; \| u^\varepsilon(0) - v^\varepsilon(0) \|_{L^2}^2 \; + \; 2\,\mathrm{Re} \int_{[0,t]\times\mathbb{R}^d} h^\varepsilon\cdot(\overline{v^\varepsilon} - \overline{u^\varepsilon})\, dt\, dx\,. \tag{6.3.4}$$

b) Set

$$\underline{w} \; := \; \underline{f} \; + \; \underline{\mathcal{E}}(\underline{u},\mathcal{U}_*)\,, \qquad \mathcal{W}_k \; := \; P_k\big(\mathcal{F}_k \; + \; \mathcal{E}_k(\underline{u},\mathcal{U}_*)\big)\,, \tag{6.3.5}$$

$$g^\varepsilon \; := \; h^\varepsilon \; - \; \underline{w} \; - \; \sum_k J_k^\varepsilon(\mathcal{W}_k)\,. \tag{6.3.6}$$

Then, Propositions 5.2.2, Corollary 5.2.4 and Proposition 6.2.4 imply that g^ε is a bounded family in $L^{1+1/p}$ which has no propagated oscillations for L.

Proposition 4.3.3 implies that the weak limit of v^ε is $\underline{u}$, v^ε is pure and the profiles are $\mathcal{U}_k$. In particular, $u^\varepsilon - v^\varepsilon$ is bounded in L^{1+p}, converges weakly to 0, is pure, and all its profiles vanish. Thus, since $\underline{w} \in L^{1+1/p}$,

$$\lim_{\varepsilon\to 0} \int_{[0,t]\times\mathbb{R}^d} \underline{w}\cdot(\overline{v^\varepsilon} \; - \; \overline{u^\varepsilon})\; dt\, dx \; = \; 0\,. \tag{6.3.7}$$

Next, since $\mathcal{W}_k \in \mathcal{L}_k^{1+1/p}$ and the profiles of $u^\varepsilon - v^\varepsilon$ vanish,

$$\lim_{\varepsilon\to 0} \int_{[0,t]\times\mathbb{R}^d} J_k^\varepsilon(\mathcal{W}_k)\cdot(\overline{v^\varepsilon} \; - \; \overline{u^\varepsilon})\; dt\, dx \; = \; 0\,. \tag{6.3.8}$$

Finally, since u^ε and v^ε are bounded families in L^{1+p} such that Lu^ε and Lv^ε are bounded in $L^{1+1/p}$, and since g^ε has no propagated oscillations for L, one has

$$\lim_{\varepsilon\to 0} \int_{[0,t]\times\mathbb{R}^d} g^\varepsilon\cdot(\overline{v^\varepsilon} \; - \; \overline{u^\varepsilon})\; dt\, dx \; = \; 0\,. \tag{6.3.9}$$

Therefore,

$$\lim_{\varepsilon\to 0}\int_{[0,t]\times\mathbb{R}^d} h^\varepsilon\cdot(\overline{v^\varepsilon}-\overline{u^\varepsilon})\,dt\,dx \;=\; 0\,. \tag{6.3.11}$$

c) The Cauchy data for u^ε and v^ε satisfy (2.5.1) and (6.2.12) respectively. Therefore,

$$\lim_{\varepsilon\to 0}\;\|\,u^\varepsilon(0)-v^\varepsilon(0)\,\|^2_{L^2} \;=\; 0 \tag{6.3.12}$$

This shows that the right hand side of (6.3.4) tends to 0 as ε tends to 0, implying the Proposition.

COROLLARY 6.3.2. *The weak limit $\underline{u}$ and the profiles $\mathcal{U}_k$ satisfy the equations*

$$L\underline{u} \;+\; \underline{\mathcal{E}}(\underline{u},\mathcal{U}_*) \;=\; 0\,, \qquad \text{on } [0,\infty[\times\mathbb{R}^d\,, \tag{6.3.13}$$

$$P_k\mathcal{U}_k \;=\; \mathcal{U}_k \quad\text{and}\quad \partial_t\mathcal{U}_k \;+\; P_k\,\mathcal{E}_k(\underline{u},\mathcal{U}_*) \;=\; 0\,, \quad \text{on } (G\backslash\mathcal{S}_k^+(p+1))\times\mathbb{T}\,, \tag{6.3.14}$$

and

$$\underline{u}|_{t=0} \;=\; \underline{u}_0 \quad \text{on } \mathbb{R}^d\,, \qquad \mathcal{U}_k|_{t=0} \;=\; P_k\mathcal{U}_0 \quad \text{on } \omega\,. \tag{6.3.15}$$

Proof. Propositions 6.3.1 and 5.2.2 imply that

$$f^\varepsilon \;\sim\; f(u^\varepsilon) \;=\; \underline{\mathcal{E}}(\underline{u},\mathcal{U}_*) \;+\; \sum_k \tilde{J}_k^\varepsilon(\mathcal{E}_k(\underline{u},\mathcal{U}_*)) \;+\; g^\varepsilon \tag{6.3.16}$$

where g^ε is bounded in $L^{1+1/p}$, converges weakly to 0 and is pure with vanishing profiles. Therefore Proposition 4.3.3 implies that the weak limit $\underline{f}$ of f^ε is $\underline{\mathcal{E}}(\underline{u};\mathcal{U}_*)$ and the profiles $\mathcal{F}_k$ are equal to $\mathcal{E}_k(\underline{u},\mathcal{U}_*)$. Thus equations (6.3.13) (6.3.14) follow from Propositions 6.1.1.

6.4. Uniqueness for the profile equations.

THEOREM 6.4.1. *Equations* (6.3.13) (6.3.14) (6.3.15) *have a unique solution* $(\underline{u},\mathcal{U}_*)$ *in* $L^{p+1}([0,\infty[\times\mathbb{R}^d)\times\mathcal{L}_*^{p+1}$.

The existence of a solution follows from Corollary 6.3.2. We must prove uniqueness. It is a consequence of the dissipativity of system (6.3.13) (6.3.12), which is inherited from the dissipativity of the original equation. Remark that this dissipativity could be used to prove directly the existence of weak solutions, using the methods of [LS] [S] [Ha] for monotone operators.

LEMMA 6.4.2. *Suppose that Ω is a regular domain. Suppose that $\underline{u}$ and $\underline{v}$ belong to $C_0^\infty(\Omega)$, that $\mathcal{U}_k$ and $\mathcal{V}_k$ are trigonometric polynomials supported in $\jmath_k^{-1}(\Omega)$ and that $P_k\mathcal{V}_k = \mathcal{V}_k$. Then*

$$\begin{aligned}\int \mathcal{E}(\underline{u},\mathcal{U}_*)\cdot\overline{\underline{v}}\,dt\,dx \;+\; \sum_k \int P_k\,\mathcal{E}_k(\underline{u},\mathcal{U}_*)\cdot\overline{\mathcal{V}_k}\,dt\,dy\,d\theta \;&= \\ \int_{\Omega\times(\mathbb{T})^Z} f\Big(t,x,\mathbf{I}(\underline{u},\mathcal{U}_*)\Big)\cdot\overline{\mathbf{I}(\underline{v},\mathcal{V}_*)}\,dt\,dx\,d\theta_*\,,&\end{aligned} \tag{6.4.1}$$

where $\mathbf{I}$ *is defined in* (5.1.3).

Proof. Introduce

$$A_k(t,y) := \int_{\mathbb{T}} P_k\,\mathcal{E}_k(u,\mathcal{U}_*)\cdot\overline{\mathcal{V}_k}\,d\theta\,. \tag{6.4.2}$$

For simplicity of notation, the (t,y) dependence of the integrand on the right hand side is suppressed. With notation as in (5.1.8), we see that

$$A_k(t,y) := \Delta_k \int_{\mathbb{T}} \mathcal{F}_k(\theta)\cdot\overline{(\mathcal{H}^{m_k}\mathcal{V}_k)\,(\theta)}\,d\theta\,, \tag{6.4.3}$$

where we have used that P_k is an orthogonal projector, $P_k\mathcal{V}_k = \mathcal{V}_k$ and that $\mathcal{H}$ is unitary in $L^2(\mathbb{T})$. According to (5.1.7), $\mathcal{F}_k$ is equal to $\mathcal{F}_k^{(1)}$ plus something independent of θ. Since the mean value of $\mathcal{V}_k$ is zero, this implies that

$$A_k(t,y) := \Delta_k(t,y)\int_{\mathbb{T}} \mathcal{F}_k^{(1)}(t,y,\theta)\cdot\overline{(\mathcal{H}^{m_k}\mathcal{V}_k)\,(t,y,\theta)}\,d\theta\,. \tag{6.4.4}$$

Introduce next the components $G_{k,j}$ of $\jmath_k^{-1}(\Omega)$ and

$$a_k(t,x) \;:=\; \sum_j \Delta_k(\sigma_{k,j}(t,x))^{-2}\,A(\sigma_{k,j}(t,x))\,, \tag{6.4.5}$$

where $\sigma_{k,j}$, an inverse of $\jmath_k$, maps Ω to $G_{k,j}$. Then, with (5.1.6), we see that

$$\begin{aligned}a_k(t,x) \;:=\; &\sum_j \Delta_k(\sigma_{k,j}(t,x))^{-1}\;\times \\ &\int_{(\mathbb{T})^Z} f\big(t,x,\mathbf{I}(\underline{u},\mathcal{U}_*)\,(t,x,\theta_*)\big)\cdot\overline{(\mathcal{H}^{m_{k,j}}\mathcal{V}_k)\,(\sigma_{k,j}(t,x),\theta_{k,j})}\,d\theta_*\,,\end{aligned} \tag{6.4.6}$$

Introduce

$$a_0(t,x) \;:=\; \underline{\mathcal{E}}(\underline{u},\mathcal{U})\,(t,x)\cdot\overline{\underline{v}(t,x)}\,. \tag{6.4.7}$$

Then, (6.4.6) and the definition (5.1.3) of $\mathbf{I}(\underline{v},\mathcal{V}_*)$ imply that

$$a_0(t,x) \;+\; \sum_k a_k(t,x) \;=\; \int_{(\mathbb{T})^Z} f\Big(t,x,\mathbf{I}(\underline{u},\mathcal{U}_*)\Big)\cdot\overline{\mathbf{I}(\underline{v},\mathcal{V}_*)}\,d\theta_*\,. \tag{6.4.8}$$

On the other hand, since Δ_k^2 is the Jacobian of q_k, one has

$$\int_\Omega a_k(t,x)\,dt\,dx \;=\; \sum_j \int_{G_{k,j}} A_k(t,y)\,dt\,dy \;=\; \int_{G_k} A_k(t,y)\,dt\,dy\,. \tag{6.4.9}$$

With (6.4.8), this implies Lemma 6.4.2.

PROPOSITION 6.4.3 *There is a constant $c > 0$ such that for all $(\underline{u}, \mathcal{U}_*)$ and $(\underline{v}, \mathcal{V}_*)$ in $L^{p+1}([0,\infty[\times\mathbb{R}^d) \times \mathcal{L}_*^{p+1}$ which satisfy $P_k\mathcal{U}_k = \mathcal{U}_k$ and $P_k\mathcal{V}_k = \mathcal{V}_k$, one has*

$$\text{Re} \int \Big(\underline{\mathcal{E}}(\underline{u}, \mathcal{U}_*) - \underline{\mathcal{E}}(\underline{v}, \mathcal{V}_*)\Big) \cdot (\overline{\underline{u}} - \overline{\underline{v}})\; dt\, dx \; +$$

$$\text{Re} \sum_k \int \Big(P_k\, \mathcal{E}_k(u, \mathcal{U}_*) - P_k\mathcal{E}_k(\underline{v}, \mathcal{V}_*)\Big) \cdot (\overline{\mathcal{U}_k} - \overline{\mathcal{V}_k})\; dt\, dy\, d\theta \;\geq \tag{6.4.10}$$

$$c\; \Big(\; \|\, \underline{u} - \underline{v} \,\|_{L^{p+1}}^{p+1} + \sum_k \|\, \mathcal{U}_k - \mathcal{V}_k \,\|_{\mathcal{L}_k^{p+1}}^{p+1} \;\Big)\,.$$

Proof. By density, one can assume that $\underline{u}$ and $\underline{v}$ belong to $C_0^\infty([0,\infty[\times\mathbb{R}^d)$ and are supported in a regular Ω. One can also suppose that the $\mathcal{U}_k$ and $\mathcal{V}_k$ are trigonometric polynomials in $\mathcal{P}_k$, supported in $G_k := \jmath_k^{-1}(\Omega)$. In this case, Lemma 6.4.2 implies that the left hand side of (6.4.10) is equal to the integral over $\Omega \times (\mathbb{T})^Z$ of

$$\text{Re}\, \Big(f(t,x,\mathbf{I}(\underline{u},\mathcal{U}_*)) \;-\; f(t,x,\mathbf{I}(\underline{v},\mathcal{V}_*))\Big) \cdot \Big(\; \overline{\mathbf{I}(\underline{u},\mathcal{U}_*)} \;-\; \overline{\mathbf{I}(\underline{v},\mathcal{V}_*)} \;\Big)\,. \tag{6.4.11}$$

Assumption 2.1.2 implies that this is larger or equal to cM where

$$M \;:=\; \int_{\Omega\times(\mathbb{T})^Z} |\, \mathbf{I}(\underline{u},\mathcal{U}_*) \;-\; \mathbf{I}(\underline{v},\mathcal{V}_*)\, |^{1+p}\; dt dx d\theta_*\,. \tag{6.4.12}$$

By definition,

$$\mathbf{I}(\underline{u},\mathcal{U}_*)\,(t,x,\theta_*) \;=\; \underline{u}(t,x) \;+\; \sum_{(k,j)\in Z} \mathbf{U}_{k,j}(t,x,\theta_{k,j}) \tag{6.4.13}$$

where

$$\mathbf{U}_{k,j}(t,x,\theta) \;:=\; \frac{1}{\Delta_k(\sigma_{k,j}(t,x))}\; \big(\mathcal{H}^{m_{k,j}}\,\mathcal{U}_k\big)(\sigma_{k,j}(t,x),\theta)\,. \tag{6.4.14}$$

Similar expressions hold for $\mathbf{I}(\underline{v},\mathcal{V})$. Since

$$\underline{u}(t,x) - \underline{v}(t,x) \;=\; \int_{(\mathbb{T})^Z} \big(\; \mathbf{I}(\underline{u},\mathcal{U}_*) \;-\; \mathbf{I}(\underline{v},\mathcal{V}_*)\big)\; d\theta_* \tag{6.4.15}$$

one has

$$\int_\Omega |\, \underline{u}(t,x) - \underline{v}(t,x)\,|^{1+p}\, dt\, dx \;\leq\; M\,. \tag{6.4.16}$$

Similarly, since $\mathbf{U}_{k,j}(t,x,\theta_{k,j})$ is the average of $\mathbf{I}(\underline{u},\mathcal{U}_*)(t,x,\theta_*)$ with respect to all the variables $\theta_{k',j'}$ except $\theta_{k,j}$, one has

$$\int_{\Omega\times\mathbb{T}} |\, \mathbf{U}_{k,j} \;-\; \mathbf{V}_{k,j}\,|^{1+p}\; dt\, dx\, d\theta \;\leq\; M\,. \tag{6.4.17}$$

Lifting the integration to $G_{k,j}$, using the boundedness of $\mathcal{H}$ in L^{1+p} and summing on j, this implies that

$$\int_{G_k\times\mathbb{T}} \Delta_k(t,y)^{1-p}\, |\,\mathcal{U}_k \;-\; \mathcal{V}_k\,|^{1+p}\; dt\, dy\, d\theta \;\leq\; C\,M\,. \tag{6.4.18}$$

This proves (6.4.10) for smooth functions supported over Ω, and thus implies Proposition 6.4.3.

Proof of Theorem 6.4.1. Suppose that $(\underline{u}, \mathcal{U}_*)$ and $(\underline{v}, \mathcal{V}_*)$ are two solutions of (6.3.13) (6.3.14). Corollary 6.2.2 implies that

$$\begin{aligned} \| \mathcal{U}_k(t) - \mathcal{V}_k(t) \|_{L^2}^2 &= \| \mathcal{U}_k(0) - \mathcal{V}_k(0) \|_{L^2}^2 - \\ &2 \operatorname{Re} \int_{[0,t]\times\omega\times\mathbb{T}} \Big(P_k\, \mathcal{E}_k(\underline{u}, \mathcal{U}_*) - P_k \mathcal{E}_k(\underline{v}, \mathcal{V}_*) \Big) \cdot (\overline{\mathcal{U}_k} - \overline{\mathcal{V}_k})\, dt\, dy\, d\theta\,. \end{aligned} \tag{6.4.19}$$

Proposition 6.2.3 implies that

$$\begin{aligned} \| \underline{u}(t) - \underline{v}(t) \|_{L^2}^2 &= \| \underline{u}(0) - \underline{v}(0) \|_{L^2}^2 - \\ &2 \operatorname{Re} \int_{[0,t]\times\omega} \Big(\underline{\mathcal{E}}(\underline{u}, \mathcal{U}_*) - \underline{\mathcal{E}}(\underline{v}, \mathcal{V}_*) \Big) \cdot (\overline{\underline{u}} - \overline{\underline{v}})\, dt\, dy\,. \end{aligned} \tag{6.4.20}$$

Adding up, and using Proposition 6.4.3 applied to functions cut off after the time t, yields

$$\begin{aligned} &\| \underline{u}(t) - \underline{v}(t) \|_{L^2}^2 + \sum_k \| \mathcal{U}_k(t) - \mathcal{V}_k(t) \|_{L^2}^2 + \\ &c \, \Big(\| \underline{u} - \underline{v} \|_{L^{p+1}}^{p+1} + \sum_k \| \mathcal{U}_k - \mathcal{V}_k \|_{\mathcal{L}_k^{p+1}}^{p+1} \Big) \\ &\qquad \le \| \underline{u}(0) - \underline{v}(0) \|_{L^2}^2 + \sum_k \| \mathcal{U}_k(0) - \mathcal{V}_k(0) \|_{L^2}^2\,. \end{aligned} \tag{6.4.21}$$

This implies uniqueness and Theorem 6.4.1 is proved.

6.5. The main result.

Theorem 2.5.1 is included in the more precise Theorem 6.5.1. We consider an equation (2.1.1) which satisfies the hyperbolicity Assumption 2.1.1 and the dissipativity Assumption 2.1.2. The initial phase ψ satisfies Assumption 2.2.1 on $\omega \subset \mathbb{R}^d$. We also suppose that the regularity, finiteness, and nonresonance Assumptions 2.3.3, 2.2.2, and 2.2.5 are satisfied.

THEOREM 6.5.1. *Suppose that $\underline{u}_0 \in L^2(\mathbb{R}^d)$, $\mathcal{U}_0 \in L^2(\omega \times \mathbb{T})$ and $\{u_0^\varepsilon\}$ is a bounded family in $L^2(\mathbb{R}^d))$ such that*

$$u_0^\varepsilon \sim \underline{u}_0 + \tilde{J}_0^\varepsilon(\mathcal{U}_0) \qquad in \quad L^2(\mathbb{R}^d)\,. \tag{6.5.1}$$

Denote by $(\underline{u}, \mathcal{U}_k)$ the unique solution of the profile equations (6.3.13-14-15) *with*

$$\begin{aligned} &\underline{u} \in C^0\big([0,\infty[\,; L^2(\mathbb{R}^d)\big) \cap L^{1+p}([0,\infty[\times\mathbb{R}^d) \\ &\mathcal{U}_k \in C^0\big([0,\infty[\,; L^2(\omega \times \mathbb{T})\big) \cap \mathcal{L}_k^{1+p}\big([0,\infty[\times\omega \times \mathbb{T}\big)\,, \end{aligned} \tag{6.5.2}$$

and by $u^\varepsilon \in C^0([0,\infty[\,; L^2(\mathbb{R}^d)) \cap L^{1+p}([0,\infty[\times\mathbb{R}^d)$ the unique solution of (2.1.1) *with initial data u_0^ε.*

Then, the family u^ε is bounded in $C^0([0,\infty[\,; L^2(\mathbb{R}^d)) \cap L^{1+p}([0,\infty[\times\mathbb{R}^d)$ and for all $T > 0$,

$$u^\varepsilon \sim \underline{u} + \sum_k \tilde{J}_k^\varepsilon(\mathcal{U}_k) \qquad in \quad L^{1+p}([0,T[\times\mathbb{R}^d)\,. \tag{6.5.3}$$

Proof. Existence, uniqueness and boundedness of u^ε is a classical result for dissipative equations (2.1.1) (see [S], [LS], [Ha]). Existence and uniqueness of $(\underline{u}, \mathcal{U}_*)$, solutions to the profile equations, are given by Theorem 6.4.1. Consider

$$v^\varepsilon \sim \underline{u} + \sum_k \tilde{J}_k^\varepsilon(\mathcal{U}_k) \qquad \text{in} \quad L^{1+p}([0,\infty[\times\mathbb{R}^d)\,. \tag{6.5.4}$$

For example, one can choose the family constructed in Proposition 6.2.4. Then, Proposition 6.3.1 implies that any pure subsequence u^{ε_n} satisfies

$$u^{\varepsilon_n} \sim v^{\varepsilon_n} \qquad \text{in} \quad L^{1+p}([0,T[\times\mathbb{R}^d)\,. \tag{6.5.5}$$

Since any bounded sequence in L^{1+p} has at least one pure subsequence by Proposition 4.3.1, this implies that the complete family u^ε satisfies

$$u^\varepsilon \sim v^\varepsilon \qquad \text{in} \quad L^{1+p}([0,+T[\times\mathbb{R}^d)\,, \tag{6.5.6}$$

proving Theorem 6.5.1.

6.6. Absorption at supercritical points

In this paragraph, we prove Theorem 2.5.4, which show that oscillations are absorbed at supercritical points. Recall the Definition 2.5.3 of $T_k^{abs}(y)$, the first time where the k-ray from y hits a supercritical point.

THEOREM 6.6.1. *If $(\underline{u}, \mathcal{U}_*)$ is the solution of the profile equations given by Theorem 6.4.1, then $\mathcal{U}_k = 0$ a.e. for $t > T_k^{abs}(y)$.*

LEMMA 6.6.2. *There exists $c > 0$ such that for all $(\underline{v}, \mathcal{V}_*) \in L^{1+p} \times L_w^{1+p}$ satisfying $P_k\mathcal{V}_k = \mathcal{V}_k$, one has for all k and almost all (t,y),*

$$\operatorname{Re} \int_{\mathbb{T}} P_k\, \mathcal{E}_k(\underline{v}, \mathcal{V}_*) \cdot \overline{\mathcal{V}_k}\, d\theta \;\geq\; c\, \Delta_k^{1-p} \int_{\mathbb{T}} |\, \mathcal{V}_k\,|^{1+p}\, d\theta\,. \tag{6.6.1}$$

Proof. By density, one can assume that $\underline{v} \in C_0^\infty([0,\infty[\times\mathbb{R}^d)$ and that the profiles $\mathcal{V}_k \in \mathcal{P}_k$. We can also assume that $(t,x) := \jmath_k(t,y)$ belongs to a regular open set Ω. Introduce

$$A_k(t,y) \;:=\; \operatorname{Re} \int_{\mathbb{T}} P_k\, \mathcal{E}_k(\underline{v}, \mathcal{V}_*) \cdot \overline{\mathcal{V}_k}\, d\theta\,. \tag{6.6.2}$$

For simplicity, the (t,y) dependence of the right hand side is suppressed. Estimate (6.6.1) is a consequence of the strict monotonicity in Assumption 2.1.2. We now find a suitable form for A_k, in order to apply inequality (2.1.3). With notations as in (5.1.7) (5.1.8), using that $P_k\mathcal{V}_k = \mathcal{V}_k$ and that $\mathcal{H}$ is unitary, we see that

$$A_k(t,y) = \Delta_k \int_{(\mathbb{T})^2} \operatorname{Re}\, \big(\mathcal{F}_k^{(1)}(\theta) - \mathcal{F}_k^{(1)}(\theta_1)\big) \cdot \overline{\mathcal{W}(\theta)}\, d\theta\, d\theta_1\,, \tag{6.6.3}$$

with $\mathcal{W} := \mathcal{H}^{m_k}\mathcal{V}_k$. Thus

$$A_k(t,y) = \frac{\Delta_k}{2} \int_{\mathbb{T}^2} \operatorname{Re}\, (\mathcal{F}_k^{(1)}(\theta) - \mathcal{F}_k^{(1)}(\theta_1)) \cdot (\overline{\mathcal{W}(\theta)} - \overline{\mathcal{W}(\theta_1)})\, d\theta\, d\theta_1\,. \tag{6.6.4}$$

By (5.1.6), the function $\mathcal{F}_k^{(1)}$ is of the form

$$\mathcal{F}_k^{(1)}(t,y,\theta) := \int_{\mathbb{T}^{Z'}} f\big(t,x,\Delta_k^{-1}(t,y)\mathcal{W}(t,y,\theta) + \mathbf{W}(t,x,\theta')\big)\, d\theta', \tag{6.6.5}$$

where, in the right hand side of (6.6.5), $(t,x) = \jmath_k(t,y)$ and $\mathbf{W}$ is a periodic function of $\theta' \in \mathbb{T}^{Z'}$. Plugging this expression in (6.6.4) transforms the right hand side into an integral over $\mathbb{T}^2 \times \mathbb{T}^{Z'}$. Assumption 2.1.2 implies that the integrated term is larger than or equal to

$$c\,\Delta_k^2\,|\,\Delta_k^{-1}\mathcal{W}(\theta) - \Delta_k^{-1}\mathcal{W}(\theta_1)\,|^{1+p} = c\,\Delta_k^{1-p}\,|\,\mathcal{W}(\theta) - \mathcal{W}(\theta_1)\,|^{1+p}. \tag{6.6.6}$$

Thus

$$A_k(t,y) \geq c\,\Delta_k^{1-p} \int_{\mathbb{T}^2} |\,\mathcal{W}(\theta) - \mathcal{W}(\theta_1)\,|^{1+p}\, d\theta d\theta_1 . \tag{6.6.7}$$

Since the average of $\mathcal{W}$ is zero, $\mathcal{W}(\theta) = \int (\mathcal{W}(\theta) - \mathcal{W}(\theta_1))\, d\theta_1$. Thus recalling $d\theta_1$ is normalized,

$$\int_{\mathbb{T}} |\,\mathcal{W}(\theta)\,|^{p+1}\, d\theta \leq \int_{\mathbb{T}^2} |\,\mathcal{W}(\theta) - \mathcal{W}(\theta_1)\,|^{p+1}\, d\theta\, d\theta_1 . \tag{6.6.8}$$

Since $\mathcal{H}$ is bounded in $L^{p+1}(\mathbb{T})$, (6.6.7) and (6.6.8) imply (6.6.1) and the lemma is proved.

PROPOSITION 6.6.3. *Consider the solution $(\underline{u}, \mathcal{U}_*)$ of the profile equations given by Theorem* 6.4.1. *Introduce*

$$\sigma_k(t,y) := \int_{\mathbb{T}} |\,\mathcal{U}_k(t,y,\theta)\,|^2\, d\theta . \tag{6.6.9}$$

Then, $\sigma_k \in C^0([0,\infty[\,; L^1(\mathbb{R}^d))$, both $\partial_t \sigma_k$ and $\Delta(t,y)^{1-p}\,\sigma(t,y)^{(1+p)/2}$ belong to $L^1([0,\infty[\,\times \mathbb{R}^d)$, and there is a constant $c > 0$ such that

$$\partial_t \sigma_k(t,y) + c\,\Delta_k(t,y)^{1-p}\,\sigma_k(t,y)^{(1+p)/2} \leq 0 . \tag{6.6.10}$$

Note that inequality (6.6.10) holds everywhere, including neighborhoods of supercritical points. This is in contrast to the profile equation (6.3.14) which holds only on the complement of the supercritical points.

Proof. We know that $\mathcal{U}_k \in \mathcal{L}_k^{1+p}$, $P_k \mathcal{U}_k = \mathcal{U}_k$, $\mathcal{F}_k := P_k \mathcal{E}_k(\underline{u}, \mathcal{U}_*) \in \mathcal{L}_k^{1+1/p}$ and that $\partial_t \mathcal{U}_k + \mathcal{F}_k = 0$ on G_k. From Corollary 6.2.2, $\mathcal{U}_k$ is continuous in t with values in L^2, implying that

$$\sigma_k \in C^0([0,\infty[\,; L^1(\mathbb{R}^d)) . \tag{6.6.11}$$

Next we show that

$$\partial_t \sigma_k + 2\,\mathrm{Re} \int_{\mathbb{T}} \mathcal{F}_k \cdot \overline{\mathcal{U}_k}\, d\theta = 0 \tag{6.6.12}$$

in the sense of distributions on G. To prove this, approximate $\mathcal{U}_k$ and $\mathcal{F}_k$ by trigonometric polynomials $\mathcal{U}_k^n$ and $\mathcal{F}_k^n$ according to Proposition 6.2.1. The identity

(6.6.12) for the approximations is a consequence of the equation $\partial_t \mathcal{U}_k^n + \mathcal{F}_k^n = 0$. To pass to the limit, use Corollary 6.2.2 which implies that σ_k^n converges to σ_k in $C^0([0,\infty[\,;L^1(\omega))$, and Proposition 6.2.1 which implies that $\mathcal{F}_k^n \cdot \overline{\mathcal{U}_k^n}$ converges to $\mathcal{F}_k \cdot \overline{\mathcal{U}_k}$ in $L^1(G)$. In particular, this proves that $\partial_t \sigma_k \in L^1$.

Lemma 6.6.2 asserts that

$$\text{Re} \int_{\mathbb{T}} \mathcal{F}_k \cdot \overline{\mathcal{U}_k}\, d\theta \ \geq c\, \Delta(t,y)^{1-p} \int_{\mathbb{T}} |\mathcal{U}_k(t,y,\theta)\,|^{p+1}\, d\theta \,. \tag{6.6.13}$$

Finally, note that

$$\sigma_k^{(1+p)/2}(t,y) \ \leq \ \int_{\mathbb{T}} |\mathcal{U}_k(t,y,\theta)|^{1+p}\, d\theta \,. \tag{6.6.14}$$

Since $\mathcal{U}_k$ belongs to $\mathcal{L}_k^{1+p}$ this implies that $\Delta_k^{1+p}\, \sigma_k^{(1+p)/2} \in L^1$. Thus, (6.6.10) and the proposition follow from (6.6.11), (6.6.13), (6.6.14).

Proof of Theorem 6.6.1. We must show that $\mathcal{U}_k = 0$ for almost all (t,y) such that $t > T_k^{abs}$. It is sufficient to consider y in a small neighborhood ω' of $\underline{y}$, contained in ω_0. In this case, $r(y) := T_k^{abs}(y)$ is a smooth function of y and Assumption 2.3.3 implies that

$$\Delta_k(t,y) \ \approx \ |t - r(y)|^{\mu/2} \tag{6.6.15}$$

for $y \in \omega'$ and $t \in I$ a neighborhood of $\underline{t} = r(\underline{y})$. In addition, $(\underline{t}, \underline{y})$ is supercritical, which means that

$$\alpha \ := \ \mu\,(p-1)/2 \ \geq \ 1\,. \tag{6.6.16}$$

Proposition 6.6.3 implies that on $I \times \omega'$, σ_k satisfies

$$\begin{aligned} &\sigma_k \in C^0(I; L^1(\omega'))\,, \qquad \sigma_k \geq 0\,, \\ &\partial_t \sigma_k \in L^1(I \times \omega')\,, \qquad |t - r(y)|^{-\alpha}\, \sigma_k^{(1+p)/2} \in L^1(I \times \omega')\,, \\ &\partial_t \sigma_k \ + \ c\, |t - r(y)|^{-\alpha}\, \sigma_k^{(1+p)/2} \ \leq 0\,. \end{aligned} \tag{6.6.17}$$

To complete the proof of Theorem 6.6.1, it remains to show that (6.6.17) implies that $\sigma_k = 0$ for $t > r(y)$.

To prove this, one can change the time variable t to $t - r(y)$, and thus reduce the problem to the case where $r \equiv 0$. Introduce

$$g(t) \ := \ \int_{\omega'} \sigma_k(t,y)\, dy. \tag{6.6.18}$$

Then

$$g(t)^{(1+p)/2} \ \leq \ (\text{meas } \omega')^{(p-1)/2} \int_{\omega'} \sigma_k(t,y)^{(1+p)/2}\, dy. \tag{6.6.19}$$

Thus (6.6.17) implies that there exist $c > 0$ such that on $I := [-a, +a]$, g satisfies

$$\begin{aligned} &g \in C^0(I)\,, \qquad g \geq 0\,, \qquad g' \in L^1(I)\,, \qquad |t|^{-\alpha}\, g^{(1+p)/2} \in L^1(I)\,, \\ &g' \ + \ c\, |t|^{-\alpha}\, g^{(1+p)/2} \ \leq 0\,. \end{aligned} \tag{6.6.20}$$

This implies that $g' \le 0$, hence that g is nonincreasing in particular $g(t) \ge g(0)$ for $t < 0$. Since $|s|^{-\alpha}\, g(s)^{(1+p)/2}$ is integrable over $[-a, 0\,[$ and $\alpha \ge 1$, this implies that $g(0) = 0$. Therefore $g(t) = 0$ for $t > 0$, so $\sigma_k(t, y) = 0$ a.e. for $t > 0$ and Theorem 6.6.1 is proved.

Proposition 6.6.3 has another corollary. It shows that no oscillations can be created in the k-mode by nonlinear interaction, if there are no oscillations in this mode at time $t = 0$.

THEOREM 6.6.4. *Let $(\underline{u}, \mathcal{U}_*)$ be the solution of the profile equations given by Theorem 6.4.1. Suppose that $P_k(y)\,\mathcal{U}_0(y, \theta)$, the initial condition for $\mathcal{U}_k$, vanishes on $\omega' \times \mathbb{T}$. Then $\mathcal{U}_k = 0$ a.e. on $[0, \infty\,[\, \times \omega' \times \mathbb{T}$.*

Proof. Proposition 6.6.3 implies that the function

$$g(t) \;:=\; \int_{\omega'} \sigma_k(t, y)\; dy \tag{6.6.21}$$

is continuous and that its derivative is integrable and nonpositive. Thus g is nonincreasing and $g(t)$ vanishes identically if $g(0) = 0$.

6.7. Angular smoothness for profiles

One of the difficulties in our analysis is that, for profiles $\mathcal{U}_k$ which are not smooth, the expression $\mathcal{U}_k(t, y, \psi(y)/\varepsilon)$ does not make sense for fixed ε. However, if $\mathcal{U}$ is smooth in θ, the substitution is meaningful. One can prove regularity in θ by using the contractivity estimate (6.4.21) together with the fact that that the profile equations are invariant under translations in θ. The angular smoothness is described using the spaces $\mathcal{L}_k^{s,q}$ introduced in Definition 2.4.1.

THEOREM 6.7.1. *In addition to the assumptions of Theorem 6.5.1, suppose that $\partial_\theta \mathcal{U}_0 \in L^2(\omega \times \mathbb{T})$. Then the profiles $\mathcal{U}_k$ belong to $\mathcal{L}_k^{s,p+1}$ for all $s < 2/(p+1)$ and $\partial_\theta \mathcal{U}_k$ belong to $L^\infty([0, \infty[\,: L^2(\omega \times \mathbb{T}))$. In particular, the solutions u^ε satisfy*

$$u^\varepsilon \;\sim\; \underline{u} + \sum_k J_k^\varepsilon(\mathcal{U}_k) \qquad \text{in} \quad L^{p+1}([0, T] \times \mathbb{R}^d)\,. \tag{6.7.1}$$

Proof. The profile equations are invariant by translations in the θ variable. Thus $(\underline{v}, \mathcal{U}^h) \;:=\; (\,\underline{u}, \mathcal{U}(t, y, \theta + h)\,)$ is another solution. Therefore inequality (6.4.21) implies that

$$\begin{aligned} \sup_{0 \le t \le T} \sum_k \| \mathcal{U}_k(t) - \mathcal{U}_k^h(t) \|_{L^2}^2 \;+\; c \sum_k \| \mathcal{U}_k - \mathcal{U}_k^h \|_{\mathcal{L}_k^{p+1}}^{p+1} \\ \le \sum_k \| \mathcal{U}_k(0) - \mathcal{U}_k^h(0) \|_{L^2}^2\,. \end{aligned} \tag{6.7.2}$$

Since $\partial_\theta \mathcal{U}_0 \in L^2(\omega \times \mathbb{T})$, the right hand side of (6.7.2) is $O(h^2)$ and one concludes that $\partial_\theta \mathcal{U}_k \in L^\infty([0, \infty[\,; L^2(\omega \times \mathbb{T}))$ and

$$\sup_{0 \le t \le T} \left\| \frac{\partial \mathcal{U}_k(t)}{\partial \theta} \right\|_{L^2(\omega \times \mathbb{T})} \;\le\; \left\| \frac{\partial \mathcal{U}_k(0)}{\partial \theta} \right\|_{L^2(\omega \times \mathbb{T})}. \tag{6.7.3}$$

Equation (6.7.2) also implies that

$$\sup_{0<h\leq 1} \int_{[0,T]\times\omega\times\mathbb{T}} \Delta_k(t,y)^{1-p} \left| \frac{\mathcal{U}_k(t,y,\theta+h) - \mathcal{U}_k(t,y,\theta)}{h^{2/(p+1)}} \right|^{p+1} dt\, dy\, d\theta$$
$$\leq C \left\| \frac{\partial \mathcal{U}_k(0)}{\partial \theta} \right\|_{L^2(\omega\times\mathbb{T})}.$$

Thus, for all $s < 2/(p+1)$,

(6.7.4)
$$\int_0^1 \int_{[0,T]\times\omega\times\mathbb{T}} \Delta_k(t,y)^{1-p} \left| \frac{\mathcal{U}_k(t,y,\theta+h) - \mathcal{U}_k(t,y,\theta)}{h^s} \right|^{p+1} dt\, dy\, d\theta\, \frac{dh}{h}$$
$$\leq C \left\| \frac{\partial \mathcal{U}_k(0)}{\partial \theta} \right\|_{L^2(\omega\times\mathbb{T})}.$$

which implies that $\mathcal{U}_k \in \mathcal{L}_k^{s,p+1}$. Proposition 4.2.1 then implies that $J_k^\varepsilon(\mathcal{U}_k)$ is defined for each ε and that and is a representative of $\tilde{J}_k^\varepsilon(\mathcal{U}_k)$ and (6.7.1) is equivalent to (6.5.3).

7. Additional results and remarks

7.1. The wave equation

With minor changes which we now explain, our results apply to the semilinear wave equation

$$\Box v \;+\; g(t,x,\partial_t v) \;=\; 0\,. \tag{7.1.1}$$

ASSUMPTION 7.1.1. *$g \in C^1([0,\infty[\times\mathbb{R}^d \times \mathbb{C}\,;\,\mathbb{C})$, $g(t,x,0)=0$ and there are constants $p\in[1,\infty[$ and $0<c<C<\infty$ such that for all $(t,x)\in\mathbb{R}^{1+d}$ and all v and v' in $\mathbb{C}$,*

$$|\,g(t,x,v)\,| \le \; C\,|\,v\,|^p\,, \qquad |\,\partial_{v,\overline{v}}g(t,x,v)\,| \le \; C\,|\,v\,|^{p-1}\,,$$

and

$$\mathrm{Re}\,\big(g(t,x,v)-g(t,x,v')\big)\,(\overline{v}-\overline{v}') \;\ge\; c\,|\,v\,-\,v'\,|^{p+1}\,.$$

Consider oscillatory Cauchy data

$$v^\varepsilon|_{t=0} = \underline{h}_0(x) + \varepsilon\,\mathbf{H}_0(x,\psi(x)/\varepsilon)\,, \qquad \partial_t v^\varepsilon|_{t=0} = \underline{h}_1(x) + \mathbf{H}_1(x,\psi(x)/\varepsilon)\,, \tag{7.1.2}$$

where ψ is a smooth function with nonvanishing differential on $\omega\subset\mathbb{R}^d$, the initial profiles $\mathbf{H}_0(x,\theta)$ and $\mathbf{H}_1(x,\theta)$ are smooth 2π-periodic in θ with mean 0 which are compactly supported in $\omega\times\mathbb{T}$, and the initial means $\underline{h}_0,\underline{h}_1$ belong to $C_0^\infty(\mathbb{R}^d)$.

Introduce $u := (\partial_t v, \partial_x v) := (u^0,u^1,\dots,u^d)$. Then (7.1.1) (7.1.2) take the form

$$Lu \;:=\; \partial_t u \;-\; A(\partial_x)\,u \;+\; f(t,x,u) \;=\; 0\,, \tag{7.1.3}$$

$$u^\varepsilon_{|}t=0 \;=\; \underline{u}_0(x) \;+\; \mathbf{U}_0(x,\psi(x)/\varepsilon) \;+\; O(\varepsilon)\,, \tag{7.1.4}$$

with

$$A(\xi) \;:=\; \begin{pmatrix} 0 & {}^t\xi \\ \xi & 0 \end{pmatrix}, \qquad f(t,x,u) \;:=\; \begin{pmatrix} g(t,x,u_0,) \\ 0 \end{pmatrix}, \tag{7.1.5}$$

and

$$\underline{u}_0 := (\underline{h}_1, \partial_x\underline{h}_0)\,, \qquad \mathbf{U}_0 := (\mathbf{H}_1, \partial_x\psi\,\partial_\theta\mathbf{H}_0)\,. \tag{7.1.6}$$

Equation (7.1.3) is of the form (2.1.1). The Assumption 2.1.1 of hyperbolicity and constant multiplicity is satisfied, but the dissipativity Assumption 2.1.2 is not satisfied. However, f depends only on u^0 and the energy methods for (7.1.1) involve scalar products $f(t,x,u)\cdot\overline{u} = g(t,x,u^0)\cdot\overline{u^0}$, suggesting that Assumption 7.1.1 is sufficient. More generally, our methods apply to systems (2.1.1) when the nonlinearity f satisfies

ASSUMPTION 7.1.2. *There is a subspace* $E \subset \mathbb{C}^N$, *and a projector* p_E *from* $\mathbb{C}^N$ *to* E, *and a mapping* $g(t,x,v)$ *from* $[0,\infty[\times\mathbb{R}^d\times E$ *to* E, *which satisfies Assumption* 2.1.2 *on* $[0,+\infty[\times\mathbb{R}^d \times E$, *and in addition*

$$f(t,x,u) := g(t,x,p_E u)\,. \tag{7.1.7}$$

In the reduction of the wave equation above, E is the space of $u=(u^0,u')\in\mathbb{C}\times\mathbb{C}^d$ such that $u'=0$. Assumption 7.1.2 guarantees the global existence and uniqueness of solutions to (2.1.1), $u\in C^0([0,\infty[\,;L^2(\mathbb{R}^d))$, with $p_E u\in L^{p+1}([0,\infty[\times\mathbb{R}^d)$.

The reduction of $\Box$ to a first order system introduces the artificial eigenvalue $\tau=0$ and thus the possibility of artificial resonances. To circumvent this difficulty, we note, as in [JMR 3], that the system (7.1.3) satisfies

ASSUMPTION 7.1.3. *L satisfies Assumption* 2.1.1 *and there is a subset* $K\subset\{1,\dots,k_0\}$ *so that the subvariety*

$$\Gamma := \bigcup_{k\in K}\{\tau=\lambda_k(\xi)\}\subset \text{char }L$$

satisfies the following condition. There are linear mappings $R(\tau,\xi)$ *from* E *to* $\mathbb{C}^N$, *which depends smoothly on* $(\tau,\xi)\in\mathbb{R}^{1+d}\backslash(\Gamma\cup\{0\})$ *and*

$$\forall(\tau,\xi)\notin\Gamma\cup\{0\}\,,\quad \forall u\in E\,,\quad L(\tau,\xi)\,R(\tau,\xi)\,u \;=\; u\,. \tag{7.1.8}$$

For the system (7.1.3) this assumption is satisfied with $\Gamma:=\{\tau^2=|\xi|^2\}$ and

$$R(\tau,\xi)\;:\;\begin{pmatrix} b\\ 0\end{pmatrix}\;\mapsto\;\frac{b}{\tau^2-|\xi|^2}\begin{pmatrix}\tau\\ \xi\end{pmatrix}. \tag{7.1.9}$$

Assumption 7.1.3 implies that if $\varphi\in C^\infty(\Omega)$ and $b\in C_0^\infty(\Omega)$ satisfy

$$d\varphi(t,x)\notin\Gamma \quad\text{and}\quad b(t,x)\in E \quad\text{a.e.} \tag{7.1.10}$$

then $b\,e^{i\varphi/\varepsilon}$ has no propagated oscillations for L. To prove this modification of Propositions 5.2.2 and 5.2.3, it is sufficient to replace $L^{-1}(d\varphi)$ by $R(d\varphi)$ in (5.2.13). This shows that only the resonances with $d\varphi\in\Gamma$ need be considered. In this spirit, introduce $\jmath_k$, $\mathcal{S}_k$, J_k^ε etc, only for $k\in K$. In particular the nonresonance Assumption 2.2.5 is weakened to

ASSUMPTION 7.1.4 *For all regular open* Ω, *the set of phases* $\varphi_{k,j}$, $(k,j)\in Z\subset K\times\mathbb{N}$, *given by* (2.2.14), *satisfy for all* $\alpha\in\mathbb{Z}^Z$ *with at least two nonvanishing components,*

$$\sum \alpha_{k,j}d\varphi_{k,j}\;\notin\Gamma\cup\{0\}\quad\text{a.e. on }\Omega\,. \tag{7.1.11}$$

THEOREM 7.1.5. *Consider Cauchy data of the form* (2.5.1), *with*

$$\forall k \notin K\,, \qquad P_k \mathcal{U}_0 \;=\; 0\,. \tag{7.1.12}$$

Then, there are unique $\underline{u} \in C^0([0,\infty[\,;\,L^2(\mathbb{R}^d)) \cap L^{p+1}([0,\infty[\times\mathbb{R}^d)$ *and* $\mathcal{U}_k \in \mathcal{L}_k^{p+1} \cap C^0([0,\infty[\,;\,L^2(\omega\times\mathbb{T}))$, $k \in K$, *such that*

$$u^\varepsilon \;\sim \underline{u} \,+\, \sum_{k\in K} \tilde{J}_k^\varepsilon(\mathcal{U}_k) \quad \text{in} \quad L^{p+1}([0,T[\times\mathbb{R}^d)\,. \tag{7.1.13}$$

The $\underline{u}$ *and* $\mathcal{U}_k$ *are solutions of equations, analogous to* (6.3.13-14-15), *where the operators* $\underline{\mathcal{E}}$ *and* $\mathcal{E}_k$ *only involve the Lagrangians* Λ_l *for* $l \in K$.

For the system (7.1.3), Γ is the union of $\tau = |\xi|$ and $\tau = -|\xi|$ and K has cardinality two. The phases $\varphi_{k,j}$, all satisfy the eikonal equation for $\square$

$$(\partial_t\varphi)^2 \;=\; |\partial_x\varphi|^2\,. \tag{7.1.14}$$

Assumption 7.1.11, means that over regular open sets Ω, one has

$$\Big(\sum \alpha_{k,j}\partial_t\varphi_{k,j}\Big)^2 \,-\, \Big|\sum \alpha_{k,j}\partial_x\varphi_{k,j}\Big|^2 \;\neq 0 \quad \text{a.e. on } \Omega\,. \tag{7.1.15}$$

Finally, note that the condition (7.1.12) is satisfied for Cauchy data (7.1.4). For the solutions v^ε of (7.1.1) (7.1.2), one has

$$\begin{cases} \partial_t v^\varepsilon \;\sim\; \partial_t\underline{v} \,+\, \tilde{J}_+^\varepsilon(\mathcal{B}_+) \,+\, J_-^\varepsilon(\mathcal{B}_-) \\ \partial_{x_j} v^\varepsilon \;\sim\; \partial_{x_j}\underline{v} \,+\, J_+^\varepsilon(\mathcal{B}_{+,j}) \,+\, J_-^\varepsilon(\mathcal{B}_{-,j}) \end{cases} \quad \text{in} \quad L^{p+1}([0,T[\times\mathbb{R}^d)\,, \tag{7.1.16}$$

with

$$\mathcal{B}_{\pm,j} \;=\; \pm\,\frac{\partial_{y_j}\psi}{|d\psi|}\,\mathcal{B}_\pm\,. \tag{7.1.17}$$

The profile equations are

$$\begin{cases} \square\underline{v} & +\;\; \underline{\mathcal{E}}(\partial_t\underline{u},\mathcal{B}_+,\mathcal{B}_-) \;=\; 0\,, \\ 2\,\partial_t\mathcal{B}_+ & +\;\; \mathcal{E}_+(\partial_t\underline{u},\mathcal{B}_+,\mathcal{B}_-) \;=\; 0\,, \\ 2\,\partial_t\mathcal{B}_- & +\;\; \mathcal{E}_-(\partial_t\underline{u},\mathcal{B}_+,\mathcal{B}_-) \;=\; 0\,, \end{cases} \tag{7.1.18}$$

where the operators $\underline{\mathcal{E}}$ and $\mathcal{E}_\pm$ are those associated to the nonlinearity $g(t,x,\partial_t v)$. Note that the profiles of $u^\varepsilon := (\partial_t v^\varepsilon, \partial_x v^\varepsilon)$ are $\mathcal{B}_\pm\,(1, \pm d\psi/|d\psi|)$ and the orthogonal projector on the line generated by $\ell := (1, \pm d\psi/|d\psi|)$ is $\ell\otimes\ell/2$. This explains the factor 2 in front of $\partial_t\mathcal{B}_\pm$ in (7.1.18).

The initial conditions are

$$\begin{cases} \underline{v}|_{t=0}(x) \;=\; \underline{h}_0(x)\,, \qquad \partial_t\underline{v}|_{t=0}(x) \;=\; \underline{h}_1(x) \\ \mathcal{B}_{\pm|t=0}(x,\theta) \;=\; \frac{1}{2}\left(\mathbf{H}_1(x,\theta) \;\pm\; |d\psi(x)|\;\partial_\theta\mathbf{H}_0(x,\theta)\right). \end{cases} \tag{7.1.19}$$

7.2. Lipschitzean nonlinearities and nonstrictly monotone nonlinearities

The analysis and results extend immediately to nonlinear terms

$$f(t,x,u) + h(t,x,u)$$

where f is as in §2 and the perturbation h vanishes at $u = 0$ and is globally lipschitzean in u. In this way the results include both those of [JMR 2] and [JMR 3]. More generally, one can treat nonlinear terms satisfying the three conditions,

$$c_1\,|u|^{p+1} - c_2\,|u|^2 \le \mathrm{Re}\ f(t,x,u)u\,, \qquad |f(t,x,u)| \le C_1|u|^p + C_2|u|^2\,, \tag{7.2.1}$$

and

$$\mathrm{Re}\Big(f(t,x,u) - f(t,x,v)\,,\,u - v\Big) \ge c\,|u-v|^{p+1} - c'\,|u-v|^2\,, \tag{7.2.2}$$

with positive constants c_1 and c. In this case the solutions may grow exponentially in time, but the analysis on bounded time intervals is unchanged.

The inequalities (2.1.3) or more generally (7.2.2) with $c > 0$ serve to prove that the asymptotics (6.5.3) is valid in L^{p+1}. If the monotonicity is relaxed to $c \ge 0$, the analysis still goes through but yields (6.5.3) in the weaker sense of $L^2([0,T] \times \mathbb{R}^d)$.

7.3. Nonlinear interaction and transfer of singularities above the caustic

In this section we give an example of a beam Γ_+ which focuses and is absorbed at a fold. We study the nonlinear interaction between Γ_+ and another beam Γ_-. Let $(\underline{t},\underline{x})$ be a point in the caustic of Γ_+. The focusing of Γ_+ implies that the time derivative of the profile of Γ_- which has a finite limit as $t \nearrow \underline{t}$, tends to infinity as $t \searrow \underline{t}$.

In space dimension $d = 2$, consider the semilinear wave equation

$$\Box v \;+\; (\partial_t v)^3 \;=\; 0 \tag{7.3.1}$$

with oscillatory Cauchy data

$$v^\varepsilon|_{t=0} \;=\; \varepsilon\,\mathbf{H}_0(y,\psi(y)/\varepsilon)\,, \qquad \partial_t v^\varepsilon|_{t=0} \;=\; \mathbf{H}_1(y,\psi(y)/\varepsilon)\,, \tag{7.3.2}$$

where

$$\psi(y) \;:=\; y_2 \;+\; y_1^2/2\,. \tag{7.3.3}$$

The initial profiles $\mathbf{H}_0(y,\theta)$ and $\mathbf{H}_1(y,\theta)$ are smooth and compactly supported and are 2π-periodic with mean 0 with respect to θ.

The eigenvalues are $\lambda_\pm(\xi) \;:=\; \pm|\xi|$. The mappings $\jmath_\pm$ from (2.2.4) are

$$\jmath_\pm(t,y) = (t,x) \quad \text{with} \quad x_1 = y_1 \mp t y_1/\sqrt{1+y_1^2}\,, \quad x_2 = y_2 \mp t/\sqrt{1+y_1^2}\,. \tag{7.3.4}$$

Only $\jmath_+$ leads to a caustic for positive times. The singular set (2.2.5) is

$$\mathcal{S}_+ \;:=\; \{(t,y) \in [0,\infty[\times\mathbb{R}^2 \;;\; t = s(y_1)\}\,, \tag{7.3.5}$$

where $s(y_1) := (1+y_1^2)^{3/2}$. The caustic set $\mathcal{C}_+$ is parametrized by $y \in \mathbb{R}^2$ as

$$t = (1+y_1^2)^{3/2}\,, \qquad x_1 = -y_1^3\,, \qquad x_2 = y_2 - (1+y_1^2)\,, \tag{7.3.6}$$

and

$$\mathcal{C}_+ := \{(t,x) \in [0,\infty[\times\mathbb{R}^2 \; ; \; x_1^2 = (t^{2/3}-1)^3\}. \tag{7.3.7}$$

Introduce $\mathcal{T}_+$ so that $\jmath_+^{-1}(\mathcal{C}_+) = \mathcal{S}_+ \cup \mathcal{T}_+$. $\mathcal{T}_+$ is a smooth surface parametrized by $(a,b) \in \mathbb{R}^2$ as

$$\mathcal{T}_+ := \Big\{(t,y) \in [0,+\infty[\times\mathbb{R}^2; t = (1+a^2)^{3/2}, y_1 = -a^3-a-a(1+a^2+a^4)^{1/2}, y_2 = b\Big\}$$

and there exists a smooth even convex function $\tau : \mathbb{R} \to \mathbb{R}$ such that $\tau(y_1) < s(y_1)$ for $y_1 \neq 0$, $\tau(0) = s(0)$ and

$$\mathcal{T}_+ := \{\,(t,y) \in [0,+\infty[\times\mathbb{R}^2 \; ; \; t = \tau(y_1)\,\}. \tag{7.3.8}$$

The complement $G\backslash\{\mathcal{S}_+ \cup \mathcal{T}_+\}$ has four connected components

$$\begin{aligned} G_1 &:= \{(t,y) \in [0,+\infty[\times\mathbb{R}^2 \;,\; t < \tau(y_1)\}, \\ G_2^+ &:= \{(t,y) \in [0,+\infty[\times\mathbb{R}^2 \;,\; \tau(y_1) < t < s(y_1)\,,\; y_1 > 0\}, \\ G_2^- &:= \{(t,y) \in [0,+\infty[\times\mathbb{R}^2 \;,\; \tau(y_1) < t < s(y_1)\,,\; y_1 < 0\}, \\ G_3 &:= \{(t,y) \in [0,+\infty[\times\mathbb{R}^2 \;,\; t > s(y_1)\}. \end{aligned} \tag{7.3.9}$$

Define

$$G_2 := G_2^+ \cup G_2^- . \tag{7.3.10}$$

LEMMA 7.3.1. *i) If $(t,y) \in G_1$, then (t,y) is the unique preimage of $\jmath_+(t,y)$.*

ii) If $(t,y) \in G_2$, then $\jmath_+^{-1}(\jmath_+(t,y))$ has three elements, (t,y), $(t,y') \in G_2$ and $(t,y'') \in G_3$. This defines a mapping $\rho : (t,y) \to (t,y')$ from G_2 into itself. Then $\rho \circ \rho = Id$, ρ is C^∞ on G_2, $\rho(G_2^+) = G_2^-$ and $\rho(G_2^-) = G_2^+$. Moreover, ρ extends continuously to $\overline{G_2}$ so that $\rho(\mathcal{T}_+) = \mathcal{S}_+$, $\rho(\mathcal{S}_+) = \mathcal{T}_+$.

Suppose $(t,y) \in G_2^-$. We have $(t,y') = \rho(t,y)$ if y' is the unique solution of

$$y'(1 - t(1+y'^2)^{-1/2}) = y(1 - t(1+y^2)^{-1/2}) \tag{7.3.11}$$

such that $(t,y') \in G_2^+$.

With the notations and results of §7.1, the asymptotics of $\partial_t v^\varepsilon$ are described in (7.1.16) and (7.1.18). Since the nonlinearity $\partial_t u^3$ is an odd function of $\partial_t u$, it follows that, if $\mathbf{H}_0$ is odd in θ and $\mathbf{H}_1$ is even in θ, then the average $\underline{v}$ vanishes. One has

$$\partial_t v^\varepsilon \sim J_+^\varepsilon(\mathcal{B}_+) + J_-^\varepsilon(\mathcal{B}_-) \tag{7.3.12}$$

and the profiles are determined by the transport equations

$$\left\{ \begin{array}{rcl} 2\,\partial_t\mathcal{B}_+ + \mathcal{E}_+(\mathcal{B}_+,\mathcal{B}_-) &=& 0, \\ 2\,\partial_t\mathcal{B}_- + \mathcal{E}_-(\mathcal{B}_+,\mathcal{B}_-) &=& 0, \end{array} \right. \tag{7.3.13}$$

with initial conditions

$$\left\{ \begin{array}{rcl} \mathcal{B}_+|_{t=0}(y,\theta) &=& \frac{1}{2}\Big(\mathbf{H}_1(y,\theta) + |d\psi(y)|\,\partial_\theta\mathbf{H}_0(y,\theta)\Big), \\ \mathcal{B}_-|_{t=0}(y,\theta) &=& \frac{1}{2}\Big(\mathbf{H}_1(y,\theta) - |d\psi(y)|\,\partial_\theta\mathbf{H}_0(y,\theta)\Big). \end{array} \right. \tag{7.3.14}$$

If, in addition, $\mathbf{H}_1 = |d\psi|\,\partial_\theta \mathbf{H}_0$, then Theorem 6.6.4 implies that $\mathcal{B}_-$ vanishes identically. Thus

$$\partial_t v^\varepsilon \sim J_+^\varepsilon(\mathcal{B}_+), \qquad \text{with} \qquad 2\,\partial_t \mathcal{B}_+ + \mathcal{E}_+(\mathcal{B}_+) = 0. \tag{7.3.15}$$

As mentioned in section 2.6, the exponent $p = 3$ is always supercritical, and therefore Theorem 6.6.1 implies that $\mathcal{B}_+ = 0$ on G_3.

We now describe the interaction term $\mathcal{E}_+(\mathcal{B}_+)$. On $G_1 \times \mathbb{T}$, Lemma 7.3.1 *i)* implies that

$$\mathcal{E}_+(\mathcal{B}_+)\,(t,y,\theta) = \frac{1}{2}\,\frac{1}{\Delta_+(t,y)^2}\,\mathcal{B}_+(t,y,\theta)^3, \tag{7.3.16}$$

where $\Delta_+(t,y)^2 := \big|\,1 - t/s(y_1)\,\big|$.

When $(t,y) \in G_2$, the general description of $\mathcal{E}_+$, as given in (5.1.8), involves the three preimages of $\jmath_+(t,y)$. Since $\mathcal{B}_+ = 0$ on G_3, only the preimages in G_2 have to be considered, that is (t,y) and $(t,y') = \rho(t,y)$. To (t,y'), the construction (5.1.3) assigns a variable $\theta' \in \mathbb{T}$, and (5.1.8) shows that

$$\mathcal{E}_+(\mathcal{B}_+)\,(t,y,\theta) = \frac{1}{2}\,\Delta(t,y) \int_{\mathbb{T}} \left(\frac{\mathcal{B}_+(t,y,\theta)}{\Delta_+(t,y)} + \frac{\mathcal{B}_+(\rho(t,y),\theta')}{\Delta_+(\rho(t,y))} \right)^3 d\theta'. \tag{7.3.17}$$

This formula uses the fact that the index $m(t,y)$ is equal to 0 on G_2, so that no Hilbert transform is needed. Since $\mathcal{B}_+$ is even in θ, one obtains

$$\begin{aligned} \mathcal{E}_+(\mathcal{B}_+)\,(t,y,\theta) &= \frac{1}{2}\,\frac{\mathcal{B}_+(t,y,\theta)^3}{\Delta_+(t,y)^2} + \frac{1}{2}\,a(t,y)\,\mathcal{B}_+(t,y,\theta), \\ a(t,y) &:= \frac{1}{\Delta_+(\rho(t,y))^2} \int_{\mathbb{T}} \mathcal{B}_+(\rho(t,y),\theta')^2\,d\theta'. \end{aligned} \tag{7.3.18}$$

EXAMPLE 7.3.2 Suppose that the initial oscillations are supported in $0 < \alpha \le y_1 \le \beta < \infty$. Then, Theorem 6.6.4 implies that $\mathcal{B}_+$ is supported in the strip $\mathcal{D} := \{(t,y) \in [0,\infty[\times\mathbb{R}^2 \;:\; \alpha \le y_1 \le \beta\}$. If $(t,y) \in \mathcal{D} \cap G_2 \subset G_2^+$, then $\rho(t,y) \in G_2^-$ and $\mathcal{B}_+(\rho(t,y),\,.\,)$ vanishes identically on $\mathbb{T}$. Therefore, $a(t,y) = 0$. Combining (7.3.16) and (7.3.18) and the absorption result on G_3, we conclude that $\mathcal{B}_+$ is supported in $K \times \mathbb{T}$, $K := \{(t,y) \in [0,\infty[\times\mathbb{R}^2 : \alpha \le y_1 \le \beta,\ 0 \le t \le s(y_1)\}$ and $\mathcal{B}$ satisfies in the interior of $K \times \mathbb{T}$

$$2\,\partial_t\,\mathcal{B}_+(t,y,\theta) + \frac{\mathcal{B}_+(t,y,\theta)^3}{\Delta_+(t,y)^2} = 0. \tag{7.3.19}$$

Thus, for $t < s(y_1)$,

$$\mathcal{B}_+(t,y,\theta) = \frac{\mathcal{B}_+(0,y,\theta)}{\big(1 + \ell(t,y)\,\mathcal{B}_+(0,y,\theta)^2\big)^{1/2}}, \tag{7.3.20}$$

with

$$\ell(t,y) := \int_0^t \frac{s(y_1)}{s(y_1) - s}\,ds. \tag{7.3.21}$$

The profile $\mathcal{B}_+$ is smooth if the data are smooth and tends to 0 when $t \to s(y_1)$, in accordance with the absorption theorem. Formula (7.3.20) provides a precise rate of decay which is further analysed in (7.3.38).

Note that $\jmath_+$ is a bijection from K onto $\jmath_+(K)$. In the (t,x) space, one has

$$\partial_t v^\varepsilon(t,x) \sim \mathbf{B}(t,x,\varphi(t,x)/\varepsilon) \tag{7.3.22}$$

with $\mathbf{B}$ supported in $\jmath_+(K)\times\mathbb{T}$, given by

$$\mathbf{B}(t,x,\theta) := \frac{1}{\Delta_+(t,y)}\,\mathcal{B}_+(t,y,\theta)\,, \qquad \varphi(t,x)=\psi(y)\,, \tag{7.3.23}$$

where y is the unique solution in K of $\jmath(t,y)=(t,x)$. In contrast to the behavior of the Lagrangian profile $\mathcal{B}$, the amplitude of $\mathbf{B}$ tends to infinity as (t,y) approaches $\mathcal{S}_+$. However, the rate of blow up is strictly smaller then in the linear case.

EXAMPLE 7.3.3. Suppose that the data are supported in $\{-\infty < c \le y_1 \le d < 0\} \cup \{0 < a \le y_1 \le b < +\infty\}$. Then, Theorem 6.6.4 implies that $\mathcal{B}_+$ is supported in the union of the strips $\mathcal{D}^+ := \{\,(t,y)\in[0,\infty[\times\mathbb{R}^2\ ;\ a\le y_1\le b\,\}$ and $\mathcal{D}^- := \{\,(t,y)\in[0,\infty[\times\mathbb{R}^2\ ;\ c\le y_1\le d\,\}$.

Introduce $K^+ := \{\,(t,y)\in[0,\infty[\times\mathbb{R}^2\ ;\ a\le y_1\le b,\ 0\le t\le s(y_1)\}$ and $K^- := \{\,(t,y)\in[0,\infty[\times\mathbb{R}^2\ ;\ c\le y_1\le d,\ 0\le t\le s(y_1)\}$. Then $\jmath_+$ is a bijection from $K^\pm$ onto $\jmath_+(K^\pm)$. One has

$$\partial_t v^\varepsilon(t,x) \sim \mathbf{B}^+(t,x,\varphi^+(t,x)/\varepsilon) + \mathbf{B}^-(t,x,\varphi^-(t,x)/\varepsilon) \tag{7.3.24}$$

with $\mathbf{B}^\pm$ supported in $\jmath_+(K^\pm)\times\mathbb{T}$, given by

$$\mathbf{B}^\pm(t,x,\theta) := \frac{1}{\Delta_+(t,y)}\,\mathcal{B}_+(t,y,\theta)\,, \qquad \varphi^\pm(t,x)=\psi(y)\,, \tag{7.3.25}$$

where y is the unique solution in $K^\pm$ of $\jmath_+(t,y)=(t,x)$. In this case, $\partial_t v^\varepsilon$ is the superposition of two wave trains, which interact on $\jmath_+(K^+)\cap\jmath_+(K^-)$. Each wave train focuses, $\mathbf{B}^+$ at $\jmath_+(\mathcal{S}_+\cap K^+)$ and $\mathbf{B}^-$ at $\jmath_+(\mathcal{S}_+\cap K^-)$.

When one follows a ray starting at $(0,y)\in K^-$, for $t<\tau(y_1)$, the point (t,y) remains in G_1, and $\jmath(t,y)\notin\jmath(K^+)$. There is no interaction, and $\mathcal{E}_+(\mathcal{B}_+)$ is given by (7.3.16).

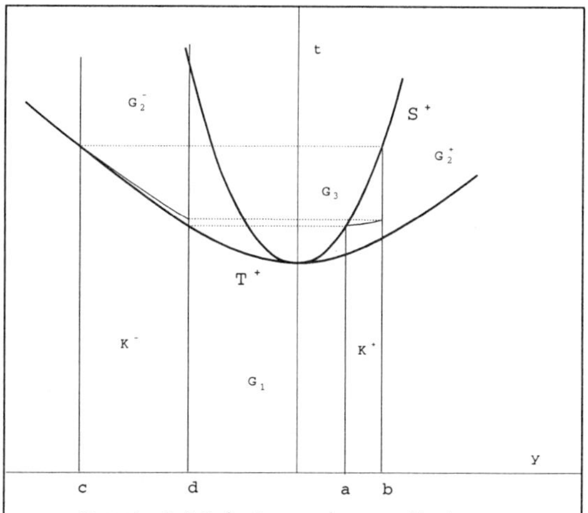

Exemple 7.3.3 in Lagrangian coordinates

Figure 7.1

When the ray reaches $\mathcal{T}_+$, that is when $(t,y) \in G_2^-$, the interaction can start since $\jmath(t,y) = \jmath(\rho(t,y))$. See Figure 7.1, where, for simplicity, it is assumed that $\jmath_+$ maps $(\mathcal{D}^+ \cap G_2)$ onto $(\mathcal{D}^- \cap G_2)$ so that, on the base, the intersection of the two beams is exactly the interaction area we want to discuss, and is wide enough to contain all the focal points of the beam corresponding to $\mathcal{D}^+$ but no more. Note that $\rho(t,y) \in \mathcal{S}_+$ when $(t,y) \in \mathcal{T}_+$, and therefore a singularity is expected in the coefficient a. Equation (7.3.18) suggests that this singularity may jump to the solution $\mathcal{B}_+$. In other words, when one follows a ray for $\mathbf{B}^-$, the interaction with $\mathbf{B}^+$ starts when one reaches the caustic of $\mathbf{B}^+$, and the singularity of $\mathbf{B}^+$ at the caustic may induce a singularity in $\mathbf{B}_-$. We now analyse this possibility in detail.

PROPOSITION 7.3.4. *i) There is a constant C such that for all $(t,y) \in K^+ \cup K^-$ and $(t,y') := \rho(t,y)$, one has*

$$\frac{1}{C}\ |t-\tau(y_1)|^{1/2} \ \le\ |t - s(y_1')|\ \le\ C\ \ |t-\tau(y_1)|^{1/2}. \tag{7.3.26}$$

ii) With a defined in (7.3.18), *there is a constant C such that for all $(t,y) \in (K^+ \cup K^-) \cap G_2$ one has*

$$|a(t,y)|\ \le\ C\ \big|t-\tau(y_1)\big|^{-1/2}. \tag{7.3.27}$$

Proof. For $(t,y) \in \overline{G_2} \cap K_\pm$ the function $t-\tau(y_1)$ vanishes only when $\rho(t,y) \in \mathcal{S}_+$, that is when $(t,y) \in \mathcal{T}_+$. Since $K^\pm$ are compact it is sufficient to prove (7.3.26) near such points. At these points, $\jmath_+$ is a local diffeomorphism near $\mathcal{T}_+$ and has a fold along $\mathcal{S}_+$. Since $\jmath_+(\mathcal{T}_+) = \jmath_+(\mathcal{S}_+)$, the estimate follows.

Proposition 6.6.3 implies that

$$\partial_t \int_{\mathbb{T}} |\mathcal{B}_+(t,y,\theta)|^2\ d\theta\ \le 0\,. \tag{7.3.28}$$

Since the data are assumed to be smooth, this implies that the integral is uniformly bounded on $[0,\infty[\times\mathbb{R}^2$.

Since $\Delta(\rho(t,y)) = |t - s(y_1')|^{1/2}$, (7.3.26) and the definition (7.3.18) of a, imply (7.3.27).

We now solve the profile equations. The solution is supported in $K \times \mathbb{T}$, with $K := K^+ \cup K^-$. On $(K \cap G_1) \times \mathbb{T}$, the equation is

$$2\ \partial_t \mathcal{B}_+\ +\ \frac{1}{\Delta_+(t,y)^2}\ \mathcal{B}_+(t,y,\theta)^3\ =\ 0\,, \tag{7.3.29}$$

where $\Delta_+(t,y)^2 := \big|\,1 - t/s(y_1)\,\big|$. The explicit solution is given by (7.3.20). Note that $\mathcal{B}_+$ is smooth on $K \cap G_1$, up to the boundary $K \cap \mathcal{T}_+$. Denote by

$$\mathcal{V}_0(y,\theta)\ :=\ \mathcal{B}_+(\tau(y_1),y,\theta)\,, \tag{7.3.30}$$

the trace of $\mathcal{B}_+$ on $\mathcal{T}_+$

On $(K \cap G_2) \times \mathbb{T}$, the equation is

$$2\,\partial_t \mathcal{B}_+(t,y,\theta) \;+\; \frac{\mathcal{B}_+(t,y,\theta)^3}{\Delta_+(t,y)^2} \;+\; a(t,y)\,\mathcal{B}_+(t,y,\theta) \;=\; 0\,, \tag{7.3.31}$$

where a is defined by (7.3.18). Introduce

$$A(t,y) \;:=\; \frac{1}{2}\int_{\tau(y_1)}^{t} a(s,y)\,ds\,. \tag{7.3.32}$$

By Proposition 7.3.4, A is bounded on $G_2 \cap K$ and vanishes on $\mathcal{T}_+$. Then

$$\mathcal{V}(t,y,\theta) \;:=\; e^{A(t,y)}\,\mathcal{B}_+(t,y,\theta) \tag{7.3.33}$$

satisfies

$$2\,\partial_t \mathcal{V}(t,y,\theta) \;+\; \frac{e^{-2A(t,y)}}{\Delta(t,y)^2}\;\mathcal{V}(t,y,\theta)^3 = 0\,, \qquad \mathcal{V}(\tau(y_1),y,\theta) = \mathcal{V}_0(y,\theta). \tag{7.3.34}$$

Since $\mathcal{B}_+$ is continuous across $\mathcal{T}_+$, one has

$$\mathcal{V}(t,y,\theta) \;=\; \frac{\mathcal{V}_0(y,\theta)}{\big(1 \,+\, h(t,y)\,\mathcal{V}_0(y,\theta)^2\big)^{1/2}}\,, \tag{7.3.35}$$

with

$$h(t,y) \;:=\; \int_{\tau(y_1)}^{t} \frac{e^{-2A(s,y)}}{\Delta(s,y)^2}\,ds\,. \tag{7.3.36}$$

Since A is bounded on $K \cap G_2$ and $\Delta(t,y)^2 = 1 - t/s(y_1)$, we see that

$$h(t,y) \;\approx\; -\,\ln(s(y_1) - t) \qquad \text{as} \qquad (t,y) \to \mathcal{S}_+ \cap K\,. \tag{7.3.37}$$

This implies that if $\mathcal{V}_0(y,\theta) \neq 0$ a.e. on $\mathbb{T}$, then

$$\mathcal{V}(t,y,\theta) \;\approx\; \Big|\ln(s(y_1) - t)\Big|^{-1/2} \quad \text{as} \quad t \to s(y_1), \quad (t,y) \in K^+. \tag{7.3.38}$$

In particular $\mathcal{V} \to 0$. But (7.3.38) is more precise. It implies that

$$\int_{\mathbb{T}} |\mathcal{V}(t,y,\theta)|^2\,d\theta \;\approx\; \Big|\ln(s(y_1) - t)\Big|^{-1} \quad \text{as} \quad t \to s(y_1), \quad (t,y) \in K^+ \tag{7.3.39}$$

if $\mathcal{V}_0(y,\,.\,) \neq 0$ a.e. on $\mathbb{T}$. Using again that A is bounded, we see that

$$\int_{\mathbb{T}} |\mathcal{B}_+(t,y,\theta)|^2\,d\theta \;\approx\; \Big|\ln(s(y_1) - t)\Big|^{-1} \quad \text{as} \quad t \to s(y_1), \quad (t,y) \in K^+. \tag{7.3.40}$$

Proposition 7.3.4 then implies that

$$a(t,y) \approx \Big|\tau(y_1) - t\Big|^{-1/2}\Big|\ln(\tau(y_1) - t)\Big|^{-1} \quad \text{as} \quad t \to \tau(y_1), \quad (t,y) \in K^-, \tag{7.3.41}$$

for $y \in \mathcal{D}^-$ satisfying $\mathcal{B}_+(0,y',\cdot) \neq 0 \in L^2(\mathbb{T})$ where $y' \in \mathcal{D}^+$ is defined by $\rho(\tau(y_1),y) = (s(y_1'),y')$. Plugging this estimate in (7.3.31) yields the following result.

THEOREM 7.3.5. *i)*

$$\lim_{t\nearrow\tau(y_1)} \partial_t\,\mathcal{B}_+(t,y,\theta) \;=\; -\,\frac{1}{2}\,\frac{\mathcal{B}_+(\tau(y_1),y,\theta)^3}{\Delta_+(\tau(y_1),y)^2}\,. \tag{7.3.42}$$

ii) If $\mathcal{B}_0(y,\theta)\neq 0$,

$$\lim_{t\searrow\tau(y_1)} \partial_t\,\mathcal{B}_+(t,y,\theta) \;=\; -\,\operatorname{sign}(\mathcal{B}_+(0,y,\theta))\,\infty\,. \tag{7.3.43}$$

Moreover, the quotient of $|\,\partial_t\,\mathcal{B}_+(t,y,\theta)\,|$ by

$$\big|\tau(y_1)-t\big|^{-1/2}\big|\ln(\tau(y_1)-t)\big|^{-1}$$

is bounded from above and from below by positive numbers, as $t\searrow\tau(y_1)$.

For globally Lipschitzean nonlinearities, we observed jumps for $\partial_t\mathcal{B}_+$ on $\mathcal{T}_+$. In particular, $\mathcal{B}_+$ was Lipschitzean in t. The singularity (7.3.43) is much stronger, since $\mathcal{B}_+$ belongs to no C^α, with $\alpha > 1/2$. Note that $\partial_t\,|\mathcal{B}_+(t,y,\theta)\,|^2 \to -\infty$ as $t\to\tau(y_1)$ from above, which means that the rate of dissipation increases drastically when one crosses $\mathcal{T}_+$.

References

[A] G. Allaire, Homogeneization and two-scale analysis, S.I.A.M. J. Math. Anal., 23, 1992, 1482-1518.

[CDMM] M. Cowling, S. Disney, G. Mauceri and D. Müller, Damping oscillatory integrals,, Invent. Math., 101, 1990, 237-260.

[Du] J.J. Duistermaat, Oscillatory integrals, Lagrangian immersions and unfolding of singularities, Comm. Pure Appl. Math. 27,1974, 207-281.

[DH] J.J. Duistermaat and L. Hörmander, Fourier integral operators II, Acta. Math., 128, 1972, 183-269.

[E] W. E, Homogeneization of linear and nonlinear transport equations, Comm. Pure Appl. Math, 45, 1992, 301-326.

[EG] L.Evans, R.Gariepy , Measure theory and fine properties of functions, Studies in Advanced Mathematic. Boca Raton, CRC Press, 1992.

[ES] W. E, D. Serre, Correctors for the homogeneization of conservation laws with oscillatory forcing terms, Asymptotic Analysis, 5, 1992, 311-316.

[Ha] A. Haraux, *Semilinear Hyperbolic Problems in Bounded Domains*, Mathematical Reports, vol. 3, Part 1, Harwood Academic Publishers, 1987.

[Hö 1] L. Hörmander, Fourier integral operators I, Acta. Math., 127, 1971, 79-183.

[Hö 2] L. Hörmander, *The Analysis of Linear Partial Differential Operators*, Springer-Verlag, 1991.

[HMR] J. Hunter, A. Majda and R. Rosales, Resonantly interacting weakly nonlinear hyperbolic waves II: several space variables, Stud. Appl. Math., 75,1986, 187-226.

[JMR 1] J.-L. Joly, G. Métivier and J. Rauch, Coherent and focusing multidimensional nonlinear geometric optics, Annales Scientifique de l'École Normale Supérieure, 28, 1995, 51-113.

[JMR 2] J.-L. Joly, G. Métivier and J. Rauch, Focusing at a point and absorption of nonlinear oscillations, Trans. Amer. Math. Soc., 347, 10, 1995, 3921-3969.

[JMR 3] J.-L. Joly, G. Métivier and J. Rauch, Nonlinear oscillations beyond caustics, Comm. Pure Appl. Math., Vol XLIX, 5, 1996, 443-527.

[JMR 4] J.-L. Joly, G. Métivier and J. Rauch, Several Recent Results in Nonlinear Geometric Optics, Partial Differential Equations and Mathematical Physics, The Danish-Swedish Analysis Seminar, Birkhauser, 1995, 181-207.

[JR] J.L. Joly and J. Rauch, Justification of multidimensional single phase semilinear geometric optics, Trans. Amer. Math. Soc., 330, 1992, 599-625.

[La] P.Lax, Asymptotic solutions of oscillatory initial value problems, Duke Math. Journal, 24, 1957, 627-645.

[L] J.-L.Lions, *Quelques Méthodes de Résolution de Problèmes aux Limites Non Linéaires*, Dunod, Gauthier-Villars, Paris, 1969.

[LS] J.-L.Lions and W. Strauss, Some nonlinear evolution equations, Bull.Soc. Math. France, 93, 1965, 43-96.

[Lu] D. Ludwig, Uniform asymptotic expansions at a caustic, Comm. Pure Appl. Math., 13, 1966, 85-114.

[McLPT] D.W. McLaughlin, G. Papanicolaou and L. Tartar, Weak limits of semi-linear hyperbolic systems with oscillating data, in *Macroscopic Modelling of Turbulent Flow*, Lecture Notes in Physics, 230, 1985, 277-298.

[N] G. Nguetseng, A general convergence result for a functional related to the theory of homogeneization, Siam J. Math. Anal., 20, 1989, 608-623.

[R 1] B.Randol, On the Fourier transform of the indicator function of a planar set, Trans. Amer. Math. Soc., 139, 1969, 271-278.

[R 2] B.Randol, On the asymptotic behavior of the Fourier transform of the indicator function of a convex set, Trans. Amer. Math. Soc., 139, 1969, 279-285.

[Ste] E. M. Stein, *Harmonic Analysis, Real Variable Methods, Orthogonality and Oscillatory Integrals*, Princeton University Press, 1993.

[Str] W. Strauss, *The Energy Method in Nonlinear Partial Differential Equations*, Notas de matematica 47, Instituto de Matematica Pura e Applicata, Rio de Janeiro 1969.

[So] C. D. Sogge, *Fourier Integrals in Classical Analysis*, Cambridge Tracts in Mathematics, 105, Cambridge University Press, 1993

[Sv] I. Svensson, Estimates for the Fourier transform of the characteristic function of a convex set, Afk. för Mat., 9, 1971, 11-22.

Jean-Luc JOLY
MAB, Université de Bordeaux I
33405 Talence, FRANCE

Guy METIVIER
IRMAR, Université de Rennes I
35042 Rennes, FRANCE

Jeffrey RAUCH
Department of Mathematics, University of Michigan
Ann Arbor 48109 MI, USA

Editorial Information

To be published in the *Memoirs*, a paper must be correct, new, nontrivial, and significant. Further, it must be well written and of interest to a substantial number of mathematicians. Piecemeal results, such as an inconclusive step toward an unproved major theorem or a minor variation on a known result, are in general not acceptable for publication. *Transactions* Editors shall solicit and encourage publication of worthy papers. Papers appearing in *Memoirs* are generally longer than those appearing in *Transactions* with which it shares an editorial committee.

As of November 31, 1999, the backlog for this journal was approximately 4 volumes. This estimate is the result of dividing the number of manuscripts for this journal in the Providence office that have not yet gone to the printer on the above date by the average number of monographs per volume over the previous twelve months, reduced by the number of issues published in four months (the time necessary for preparing an issue for the printer). (There are 6 volumes per year, each containing at least 4 numbers.)

A Copyright Transfer Agreement is required before a paper will be published in this journal. By submitting a paper to this journal, authors certify that the manuscript has not been submitted to nor is it under consideration for publication by another journal, conference proceedings, or similar publication.

Information for Authors and Editors

Memoirs are printed by photo-offset from camera copy fully prepared by the author. This means that the finished book will look exactly like the copy submitted.

The paper must contain a *descriptive title* and an *abstract* that summarizes the article in language suitable for workers in the general field (algebra, analysis, etc.). The *descriptive title* should be short, but informative; useless or vague phrases such as "some remarks about" or "concerning" should be avoided. The *abstract* should be at least one complete sentence, and at most 300 words. Included with the footnotes to the paper, there should be the 1991 *Mathematics Subject Classification* representing the primary and secondary subjects of the article. This may be followed by a list of *key words and phrases* describing the subject matter of the article and taken from it. A list of the numbers may be found in the annual index of *Mathematical Reviews*, published with the December issue starting in 1990, as well as from the electronic service e-MATH [**telnet e-MATH.ams.org** (or **telnet 130.44.1.100**). Login and password are **e-math**]. For journal abbreviations used in bibliographies, see the list of serials in the latest *Mathematical Reviews* annual index. When the manuscript is submitted, authors should supply the editor with electronic addresses if available. These will be printed after the postal address at the end of each article.

Electronically prepared papers. The AMS encourages submission of electronically prepared papers in $\mathcal{A}_{\mathcal{M}}\mathcal{S}$-TeX or $\mathcal{A}_{\mathcal{M}}\mathcal{S}$-LaTeX. The Society has prepared author packages for each AMS publication. Author packages include instructions for preparing electronic papers, the *AMS Author Handbook*, samples, and a style file that generates the particular design specifications of that publication series for both $\mathcal{A}_{\mathcal{M}}\mathcal{S}$-TeX and $\mathcal{A}_{\mathcal{M}}\mathcal{S}$-LaTeX.

Authors with FTP access may retrieve an author package from the Society's Internet node `e-MATH.ams.org` (130.44.1.100). For those without FTP access, the author package can be obtained free of charge by sending e-mail to `pub@ams.org` (Internet) or from the Publication Division, American Mathematical Society, P.O. Box 6248, Providence, RI 02940-6248. When requesting an author package, please specify $\mathcal{A}\kern-.1667em\lower.5ex\hbox{$\mathcal{M}$}\kern-.125em\mathcal{S}$-TeX or $\mathcal{A}\kern-.1667em\lower.5ex\hbox{$\mathcal{M}$}\kern-.125em\mathcal{S}$-LaTeX, Macintosh or IBM (3.5) format, and the publication in which your paper will appear. Please be sure to include your complete mailing address.

Submission of electronic files. At the time of submission, the source file(s) should be sent to the Providence office (this includes any TeX source file, any graphics files, and the DVI or PostScript file).

Before sending the source file, be sure you have proofread your paper carefully. The files you send must be the EXACT files used to generate the proof copy that was accepted for publication. For all publications, authors are required to send a printed copy of their paper, which exactly matches the copy approved for publication, along with any graphics that will appear in the paper.

TeX files may be submitted by email, FTP, or on diskette. The DVI file(s) and PostScript files should be submitted only by FTP or on diskette unless they are encoded properly to submit through e-mail. (DVI files are binary and PostScript files tend to be very large.)

Files sent by electronic mail should be addressed to the Internet address `pub-submit@ams.org`. The subject line of the message should include the publication code to identify it as a Memoir. TeX source files, DVI files, and PostScript files can be transferred over the Internet by FTP to the Internet node `e-math.ams.org` (130.44.1.100).

Electronic graphics. Figures may be submitted to the AMS in an electronic format. The AMS recommends that graphics created electronically be saved in Encapsulated PostScript (EPS) format. This includes graphics originated via a graphics application as well as scanned photographs or other computer-generated images.

If the graphics package used does not support EPS output, the graphics file should be saved in one of the standard graphics formats—such as TIFF, PICT, GIF, etc.—rather than in an application-dependent format. Graphics files submitted in an application-dependent format are not likely to be used. No matter what method was used to produce the graphic, it is necessary to provide a paper copy to the AMS.

Authors using graphics packages for the creation of electronic art should also avoid the use of any lines thinner than 0.5 points in width. Many graphics packages allow the user to specify a "hairline" for a very thin line. Hairlines often look acceptable when proofed on a typical laser printer. However, when produced on a high-resolution laser imagesetter, hairlines become nearly invisible and will be lost entirely in the final printing process.

Screens should be set to values between 15% and 85%. Screens which fall outside of this range are too light or too dark to print correctly.

Any inquiries concerning a paper that has been accepted for publication should be sent directly to the Editorial Department, American Mathematical Society, P. O. Box 6248, Providence, RI 02940-6248.

Selected Titles in This Series

(Continued from the front of this publication)

653 **L. C. Evans and W. Gangbo,** Differential equations methods for the Monge-Kantorovich mass transfer problem, 1999

652 **Arne Meurman and Mirko Primc,** Annihilating fields of standard modules of $\mathfrak{sl}(2,\mathbb{C})^{\sim}$ and combinatorial identities, 1999

651 **Lindsay N. Childs, Cornelius Greither, David J. Moss, Jim Sauerberg, and Karl Zimmermann,** Hopf algebras, polynomial formal groups, and Raynaud orders, 1998

650 **Ian M. Musson and Michel Van den Bergh,** Invariants under Tori of rings of differential operators and related topics, 1998

649 **Bernd Stellmacher and Franz Georg Timmesfeld,** Rank 3 amalgams, 1998

648 **Raúl E. Curto and Lawrence A. Fialkow,** Flat extensions of positive moment matrices: Recursively generated relations, 1998

647 **Wenxian Shen and Yingfei Yi,** Almost automorphic and almost periodic dynamics in skew-product semiflows, 1998

646 **Russell Johnson and Mahesh Nerurkar,** Controllability, stabilization, and the regulator problem for random differential systems, 1998

645 **Peter W. Bates, Kening Lu, and Chongchun Zeng,** Existence and persistence of invariant manifolds for semiflows in Banach space, 1998

644 **Michael David Weiner,** Bosonic construction of vertex operator para-algebras from symplectic affine Kac-Moody algebras, 1998

643 **Józef Dodziuk and Jay Jorgenson,** Spectral asymptotics on degenerating hyperbolic 3-manifolds, 1998

642 **Chu Wenchang,** Basic almost-poised hypergeometric series, 1998

641 **W. Bulla, F. Gesztesy, H. Holden, and G. Teschl,** Algebro-geometric quasi-periodic finite-gap solutions of the Toda and Kac-van Moerbeke hierarchies, 1998

640 **Xingde Dai and David R. Larson,** Wandering vectors for unitary systems and orthogonal wavelets, 1998

639 **Joan C. Artés, Robert E. Kooij, and Jaume Llibre,** Structurally stable quadratic vector fields, 1998

638 **Gunnar Fløystad,** Higher initial ideals of homogeneous ideals, 1998

637 **Thomáš Gedeon,** Cyclic feedback systems, 1998

636 **Ching-Chau Yu,** Nonlinear eigenvalues and analytic-hypoellipticity, 1998

635 **Magdy Assem,** On stability and endoscopic transfer of unipotent orbital integrals on p-adic symplectic groups, 1998

634 **Darrin D. Frey,** Conjugacy of Alt_5 and $\mathrm{SL}(2,5)$ subgroups of $\mathrm{E}_8(\mathbb{C})$, 1998

633 **Dikran Dikranjan and Dmitri Shakhmatov,** Algebraic structure of pseudocompact groups, 1998

632 **Shouchuan Hu and Nikolaos S. Papageorgiou,** Time-dependent subdifferential evolution inclusions and optimal control, 1998

631 **Ronnie Lee, Steven H. Weintraub, and J. William Hoffman,** The Siegel modular variety of degree two and level four/Cohomology of the Siegel modular group of degree two and level four, 1998

630 **Florin Rădulescu,** The Γ-equivariant form of the Berezin quantization of the upper half plane, 1998

629 **Richard B. Sowers,** Short-time geometry of random heat kernels, 1998

628 **Christopher K. McCord, Kenneth R. Meyer, and Quidong Wang,** The integral manifolds of the three body problem, 1998

For a complete list of titles in this series, visit the AMS Bookstore at **www.ams.org/bookstore/**.

DATE DUE
